Advance Praise for
Agile Discovery & Delivery

"Are you building the right thing? Success or failure hinges on building the right product or service with features that will add business and organizational value. *Agile Discovery and Delivery* provides an invaluable guide to juxtapose and balance discovering the right product elements in parallel with rapid delivery. Amber's practical roadmap, based on her real-life experience and thoughtful consideration, is a bright and must-read beacon for those early in their agile journey."
   - **Sanjiv Augustine**, Founder and CEO, LitheSpeed & Agile Leadership Academy

"Amber created an incredible capstone class in the CS department here at UW-Madison. Students love it! This book bottles up the magic of her class and makes it available to all. Anyone practicing the art and science of software construction will benefit from reading (and re-reading) it."
   - **Remzi Arpaci-Dusseau**, Professor & Chair, Computer Sciences, University of Wisconsin-Madison

"Every time I open the book, I find something immediately relevant to the problems I'm dealing with. It's the kind of thing you never actually stop reading, you just keep coming back for reference. It's also been kind of fun to grow with the book, I re-read the passage about discovery and testing key assumptions after 6 months and realized "hey, I did that and it worked!"
   - **Clayton Custer**, Co-Founder, EduReality, Inc.

"Amber has done a great job sharing her insights into Kanban and more over the many years of her experience applying it in the field. Using her first-hand account, you'll gain a greater understanding into some of the simple steps you can take to get started with Kanban."
  - **Joey Spooner**, Vice President for Community Development & Product Management, Kanban University

"Amber Field expertly blends the discipline of project management, the flexibility of agile methodology with the most important element of making it work for all stakeholders. I highly recommend this practical and engaging work for timely projects with teams, who after learning how to apply its principles, will enthusiastically embrace working together again."
  - **Joyce M. Sullivan**, CEO & Founder, SocMediaFin, Inc.

Agile Discovery & Delivery

Agile Discovery & Delivery

A Survival Guide for New Software
Engineers & Tech Entrepreneurs

Amber R. Field

For my kids,
Alison & Maddie

And for my husband,
Graham
whose support has helped me
reach many a life goal,
including writing this book

Contents

Introduction

My second job out of college was with a wonderful team at the National Geographic Society, and it was a fantastic experience. I worked with smart, talented people who cared about the planet and created amazing content for both the magazine and our numerous online assets. National Geographic as a workplace was unique in that you never knew who (or what!) you might run into in the hallways. For example, one morning I walked in to find several blue macaws that were to be featured in our Explorer's Symposium later that day resting on a tree limb in the lobby. I also met famous explorers in the elevator and bumped into one with his part-wolf dog in the cafeteria. In addition, we were free to watch live TEDx talks or the National Geographic competitions that we hosted on an annual basis.

However, I also experienced the worst project failure of my career at National Geographic. We spent years building a product our customers didn't use and made ourselves miserable doing it. And the kicker? We could have easily avoided such a colossal failure.

I worked at National Geographic in the early 2010s. By that time, the organization had been around for over 100 years. They had already impacted our planet tremendously and had successfully bridged the gap between scientific paper and popular press. When it came to print media, National Geographic had everything down. Each month, they delivered a new magazine based on the work of their vast network of explorers. The magazines contained beautiful colored photographs and vivid stories that captured people's imaginations, opened their eyes to new cultures, and even helped them set eyes on the Titanic once more. I looked forward to getting my copy each month and subscribed to it as a young adult.

Yet, by the time I started at National Geographic's Washington, DC office in 2011, the world no longer revolved around print media. Newspapers were dying, magazines moved online, and media outlets were struggling to install paywalls.[1] Clearly, National Geographic had to figure out how to pivot from print media to the digital age. Although they created a cable channel and several apps, they struggled to develop a cohesive digital strategy. Twenty-first Century Fox bought the team I worked with shortly after I left and later sold it to Disney in 2019.[2] The organization has since morphed into an effective digital team.

I tell the following tale with a spirit of learning and reflection, keeping in mind that most companies have had similar experiences and that we all learned an incredible amount from this project. If I hadn't gone through this experience, I wouldn't be writing this book.

This is a story about the "Membership Project," which was a large project over which I was one of the responsible project managers. National Geographic wanted to develop a platform that would connect members and explorers and where members could interact, watch videos, attend live sessions, and chat with explorers. It was a fantastic idea, and I was excited to see the project take off!

But it never did.

We worked on it for over two years and spent hours on the design. It needed to be as beautiful as the magazine and then some. To my knowledge, we didn't talk to many members, show them prototypes, or clear our ideas with the explorers. These groups were the customers for whom we were building it, and we didn't get their input.

Instead, we decided to use a contracted team of developers to create the platform. We put out a request for proposal (RFP) and hired the lowest cost partner—we'd never worked with the team before.

Executives assigned three Product Owners to develop the product requirements. Product Owners represent the customer and the business and prioritize work for technology teams. Our Product Owners rarely agreed on details, leaving the teams to figure things out for themselves. Wires were crossed. A lot of time was wasted.

Over time, the project was delayed and received more attention from the executive team. We added more engineers so we could finish faster, and we also added project scope. For example, we tried to streamline several back-end databases into one, which was a large project on its own. Managers left out of frustration, making it very difficult to manage the changes and leaving the project team to sit in a conference room and complain. My regular meetings with the team members ran long, and sometimes they devolved into full-on shouting matches. The stress was palpable. Our first attempt to deploy code took all night and still had to be reverted. I was pregnant at the time, but it was too early to share that news, so I stayed up all night with the team, practically swooning with fatigue.

We didn't deliver valuable features early and often to our customers. Instead, we worked on the platform for months before releasing a single line of code. We had success metrics that I now know were "vanity metrics"[3]—easy to gather but unsuccessful measures of the product's success. For instance, our goal was to have a million users on the site . . . they didn't have to stay there and use it. They didn't even have to be real users, to tell you the truth. This metric was too easy to fake.

The project finished over budget and very late. It did go live eventually, but executives killed it a couple of months later. After almost three years and millions of dollars of work, we were left with nothing to show for it.

What do I wish I'd known at that point in my career to avoid such a disaster? A lot of things, it turns out. For example, I wish I'd known how important it is, as an engineer, to understand my customers, as that knowledge helps us figure out what to build. Nothing is worse than spending time building the wrong thing. I wish I'd known more about collaborating on agile teams, so we could work more efficiently and sustainably together. I also wish I'd known how to discern a well-run organization from a flailing one. All our problems stemmed from the same root cause: We didn't effectively understand and employ agile software development techniques.

Today, agile software development is a mainstream practice. Eighty percent of companies create successful products using some form of agile development,[4] and hiring managers expect early-career engineers to understand the basics behind it. College hires should expect to use frameworks like Scrum or Kanban in their first jobs.

Still, as an educator and hiring manager myself, I find that most engineers don't leave college with enough agile software knowledge. Degree programs may introduce agile frameworks, but very few discuss things like discovery (how we figure out what to build) or team code branching techniques. Not only the academic world falls short in this regard; many companies apply agile frameworks unevenly across their teams. In fact, fewer than 50% of companies apply agile techniques to their entire application delivery cycle,[5] which opens up an opportunity for new engineers: Those with agile skills can teach their companies how to do things better. They can help create a fun and sustainable work environment, and they may even enjoy promotions and faster career progression than the average college hire due to the impact they're able to make at their companies.

The same holds true for tech entrepreneurs scrambling to bring their products into the world. The ability to iterate and execute quickly may mean the difference between company success and failure. Given the widespread use of agile development today, successfully deploying it has never been more important for engineers and tech entrepreneurs. This book aims to give you an overview of everything you'll need to know on the topic of agile software development, from start to finish. However, it isn't meant to overwhelm you; it's just enough knowledge to make you highly effective in your first positions.

And the National Geographic project story I told you? That's a good example of what can happen if you don't have this knowledge.

Who Should Read This Book?

If you're a software engineer or tech founder who is early in your career, this book is for you. Or, perhaps you're still a student, recently started a job with your first company, or are forging your own entrepreneurial path. If so, your education has probably prepared you well for the "coding" side of software development. You've learned languages, data structures, and algorithms, and you've likely even learned a bit about agile software development. However, what most engineering degrees lack is a deeper education around the "people" side of software development. In other words, what makes a team high-performing, and how do you contribute to that? How do you collaborate within an organization to build great software? Who are your customers? How do you figure out what to build in the first place? What makes a company fun to work for? How do you survive and thrive on your first teams?

Experienced engineers have learned answers to the questions above over time. Little by little, we uncover ideas that could have helped us long ago. My aim with this book is to give you these secrets now, while you're starting out. You can use them to find excellent organizations to work for, change the ones where you're currently employed, or build your own companies. You'll be able to step onto high-performing teams and make an impact right away. Instead of spending years learning bad habits, you'll start out with good ones. You'll even learn to survive within organizations that are still figuring things out (i.e., most of them!).

With a computer science background, you can do any type of work you'd like. You can develop cutting-edge technologies to send humans to Mars. Or, you can build products that clean up the Earth and fight climate change. You may work for social networks, large companies, non-profits, start-ups, or research projects. Software engineers can change the world, and let's face it—the world could use your help right now. This book aims to help you hit the ground running at the start of your careers, so you can make a global impact.

This book can also be useful to anyone else who works for a technology company. If you're in marketing, sales, support, or leadership and want an overview of how your product people work, read this book. A few of my readers are even simply dating or married to an engineer, and they've thanked me for the insight into what their partner is doing and for enhancing their understanding of what they're significant other is talking about at dinner.

How to Read This Book

You can read this book out of order if you want, but do please start with Chapter 1 because understanding agile principles sets the stage for everything else.

If you're starting fresh and want a great overview, read it cover to cover. I discuss the software product building process from discovery to delivery, or start to finish.

You can also jump to the section that most interests you. If you've joined a company with established products, you may want to start with "Part II: Delivery." For entrepreneurs or engineers at a start-up, go straight to Chapter 3: "Finding Product-Market Fit." This section will help you decide what to build. Within each chapter are three sections:

1. Key Concepts: The basics of what you should know about the topic.

2. Reality: How these concepts are actually executed in the real world. After all, reality is a bit messier than the theoretical perfect state.

3. Survival Tips: Simple things you can do to survive and thrive using these key concepts.

My goal is to leave you with a series of useful practices that will benefit you and your future companies. You can pick and choose whatever works best for you.

About Me

Hi! I'm Amber. I'm a leader, author, speaker, teacher, mom, runner, hiker, and a dozen other things. I hold a Computer Science degree from the University of Wisconsin-Madison, and now I teach in their Computer Science department. I also lead the product and engineering organization at Singlewire Software. I've had quite an interesting career so far as an engineer, program manager, and leader, and a lot of what I teach in the course and in this book I've picked up via my various roles at organizations small and large.

My first full-time job was as a developer at IBM. After a few years, I became a team lead, and then I moved into project and program management. During that stint, I spent two months living in Tanzania as part of IBM's first Corporate Service Corps cohort—a program patterned after the US Peace Corps. My team was composed of IBM-ers from all over the world who lent their skills to various regional non-profits. I volunteered at the African Wildlife Foundation, where they were building an eco-lodge for seven Masaai tribes living near Mount Kilimanjaro. Our task was to develop a business plan for the lodge, and the proceeds would lift the tribes out of poverty and preserve their native homeland's beautiful ecology. This was the first time I really aligned something I was passionate about (the environment) with my full-time job.

IBM Corporate Service Corps Tanzania 1 African Wildlife Foundation team with our Masaai friends in the Enduimet Wildlife Management Area

After IBM, I went to work at the National Geographic Society and moved to Washington, DC, where I discovered Scrum (an agile framework covered in Chapter 5). I also found a welcoming agile community led by my friends at Lithespeed, a local agile training firm. I ended up leading National Geographic's transition to agile software development with their help.

Along the way, I had two wonderful daughters.

In 2014, an enticing role popped up at a start-up called Opower, which used behavioral science to get people to save energy. A few years later, Oracle acquired Opower, and a lot of people left the company, including me. By that time, I was leading the Agile Program Management Office, and we were also experimenting with new innovation and product discovery techniques. (We cover some of those in Chapter 3: "Finding Product-Market Fit.")

Soon, I found a new home as Head of Operations for Capital One's Innovation Lab, where we validated early ideas and handed them over to the rest of the company when we knew they worked. I started speaking about these ideas at countless conferences on the East Coast, and I guest lectured at Harvard Business School. Later, I co-presented the keynote at the first Midwest Business Agility Conference in Columbus, OH.

By this time, I was missing my native Wisconsin, so I decided to move back to Madison in 2017. Aside from the weather, I've never looked back. (It's currently February and a chilly three degrees Fahrenheit.)

Now, I work at Singlewire Software, where I'm their Vice President of Software Development. We make InformaCast, which is a facilities-based critical event and emergency notification system. We keep people safe during all sorts of incidents, from tornados to active shooter situations. Our customers are schools, hospitals, manufacturing companies, and commercial businesses.

One fateful day, I went to an alumni happy hour at the University of Wisconsin-Madison. We had a new Computer Science Department Chair, Remzi Arpaci-Dusseau, and I asked him what the university was doing to teach students about agile software development. His answer was, "Not much, and we could use alumni like you to help us with that! Email me tomorrow about it." He jokes that he never thought he'd hear from me again, but that day was the beginning of something pretty special.

I worked with Remzi to develop the Computer Science Capstone course for UW-Madison. We partner with companies of all sizes like Capital One, Amazon, Epic Systems, American Family Insurance, Medtronic, and Last Lock. These companies develop a project idea that students work on for the entire semester on agile teams. They also provide mentors for the students. We discuss how the teams can figure out what to build, and then we learn how to build it, how to collaborate, how to pivot, and how to iterate. I layer in as many secrets, tips, and tricks as I can to help our students survive and thrive post-graduation.

This book mirrors my class. It's everything I wish I'd known coming out of college but didn't. My hope is that young software professionals everywhere (you!) learn about these concepts early. With them, you can better shape the products you work on and you'll have fun doing it.

Part I: Discovery

The word "discovery" in Part I has two meanings. The first two chapters are all about discovering agile software development's foundational principles, and we explore agile culture and what makes a great agile team. These concepts form the backbone of the agile frameworks in later chapters.

Chapter 3 is about the discovery phase in which teams determine what to build. In discovery mode, you talk to a lot of customers, find out what their problems are, and develop solutions to solve those problems.

Discovery in both cases is all about learning and laying the groundwork for future success, which is exactly what we'll be doing in Part I.

Chapter 1: An Agile Software Development Overview

Agile Overview: Key Concepts

Agile is a culture and a mindset.[1] Agile frameworks, best practices, and techniques are built on top of this culture, but any team using these frameworks must embrace the agile culture to be successful. Let's look at examples contrasting two different businesses, one of which embodied the agile culture and one which did not.

When the COVID-19 pandemic arrived in March 2020, the software industry went remote. We relied on platforms like Zoom, Microsoft Teams, and WebEx Calling to stay connected.

Zoom's founder, Eric Yuan, had been the Corporate Vice President for Cisco WebEx since Cisco bought WebEx in 2007. WebEx was Cisco's version of a software collaboration platform, and it focused on enterprise-level video conferencing. Yuan had personal experience with long distance relationship pain points because when he was in college, he traveled ten hours by train to visit his girlfriend (now wife). He wanted an easier way to communicate with her while he was away. He also knew Cisco's customers weren't happy with the WebEx platform. But, it was a minor product for a large company, and the changes he wanted to make would take time, weren't funded, or weren't possible given the existing architecture.

Yuan left Cisco in April 2011 to launch a similar video conferencing system: the Zoom platform.[2] Zoom's team emphasizes usability, design, and customer feedback. They release at least once a month to add new features and gain more customer feedback,[3] and their relentless focus on customer value catapulted Zoom to the top of the collaboration tools industry by 2021.[4] Meanwhile, Cisco WebEx languished, with long release cycles and very few updates. Zoom was able to steal a large market share by embracing an agile culture.

Cisco, for many years, used what we call the waterfall software development method; a term first coined by Winston Royce in 1970.[5] Zoom, on the other hand, used agile software development techniques. Let's take a brief look at what "waterfall" means, so we can see why the agile culture is so important.

Waterfall is a software development process that consists of several distinct stages. At its most basic, the stages are: requirement analysis, design, development, test, release, and maintenance.

THE WATERFALL PROCESS

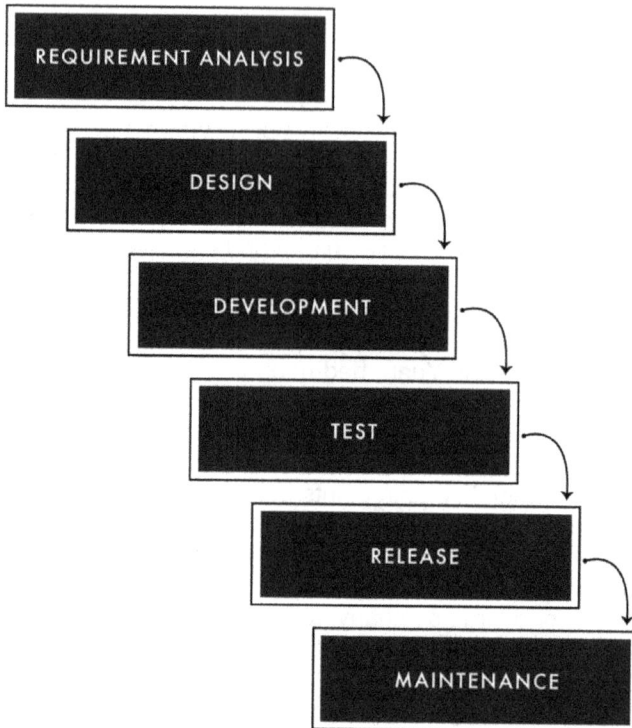

Teams work through an entire phase before embarking on the next one. In this way, work is pushed from one phase to another, thereby "falling" into the hands of the next team. The sequence is comparable to a waterfall cascading down a series of rocks—hence, the name. Usually, different teams are responsible for each phase, and the only communication between these teams is an extensive written requirements document. Many projects wouldn't ship a product for years because just getting the design and detailed requirements written might take months. Coding could take another year or two and was usually late. To meet deadlines, businesses would therefore move the release date (and face upper management's wrath) or tell the testing teams to test faster. These practices, of course, led to more customer-facing issues, as long release cycles meant expensive patches or long waits for fixes. Teams would stop feature work to fix issues, which meant the next release would be late.[6] Doesn't that sound healthy and fun?

Another problem with this strategy is that markets move faster than products can ship. During long release cycles, customer needs change, competitors enter the market, companies are acquired, and new regulations are enacted. Waterfall was the best we had at one time, but waterfall software development was a mess, resulting in most projects being late or getting canceled before release.

The first couple of projects I worked on at IBM used a "waterfall-lite" process. We spent time designing our features up-front by writing detailed use cases, and then we worked on a group of unrelated features and cobbled them into a release. We wrote automated unit tests, which was healthy, but we relied on large code merges and the integration testing phase to find most of our bugs. The integration testing phase always occurred mere weeks before our launch, and we didn't spend a lot of time talking to our users . . . at first.

In 2006, however, IBM told every team in the organization we were moving to "agile software development." Executives gave departments some documentation and told us to figure out how we wanted to run "agile" ourselves.

Our team decided to adopt one or two agile best practices every release. We focused on the practices, not the culture. At first, we held a daily fifteen-minute meeting called a "stand-up" in which we discussed the work we were doing that day. We released code every four months (which was pretty quick!), but the process still looked like waterfall. We squeezed design, implementation, testing, and integration testing serially into those four months, during which our fifteen-minute stand-ups took thirty minutes to an hour.

Still, we were learning and moving in the right direction, so we gave ourselves a pat on the back. We later formed a user group to help us determine what features to build. I was in charge of this group for a short time, and we met every few weeks to discuss customer needs. I would sometimes close their feature requests without much input if we knew we would never get to them, and we didn't show them prototypes or release previews to get feedback.

As it turns out, we were still using waterfall.

What would it have looked like if we would have actually adopted an agile culture from the beginning? For starters, our release times would have been far shorter—probably every two weeks to a month. For every release, our teams would have coded, tested, and documented our work. We would have produced and shipped working, valuable features. Our stand-ups would have actually lasted the projected fifteen minutes. We would have shared our work progress efficiently and unblocked stalled work. We would have had an organized backlog with the highest priority items at the top. Our customer feedback sessions would have been more frequent, and we would have shown them work in progress, collected their feedback, and then iterated our designs. The organization would have empowered us to make decisions quickly.

What woke most technology companies up to the fact that a far better way of working existed? A group of people wrote a literal manifesto: *The Agile Manifesto*.

The Agile Manifesto

Around the year 2000, the software development world was going through an epiphany. Many software professionals realized that the waterfall method wasn't working, so they began developing ways to improve their own projects, products, and processes. When they would talk at conferences, they realized they were trying similar things.

So, in February 2001, a group of seventeen software practitioners decided to meet in Snowbird, Utah. They all wanted to ski (according to Ron Jeffries, who joked about the meeting at a recent conference), but they also wanted to discuss the changes they were trying to make in the software industry.

The result was the *Agile Manifesto*, which has become rather famous within the technology industry[7] and still guides how we build and run healthy software teams today. The manifesto states:

Manifesto for Agile Software Development

We are uncovering better ways of developing software by doing it and helping others do it. Through this work we have come to value:

1. ***Individuals and interactions*** *over processes and tools*
2. ***Working software*** *over comprehensive documentation*
3. ***Customer collaboration*** *over contract negotiation*
4. ***Responding to change*** *over following a plan*

That is, while there is value in the items on the right, we value the items on the left more.

The manifesto also outlines twelve fantastic principles, but I've boiled them down to ten.

Agile Principles

These ten agile principles can be a powerful force for good as they help teams build great software and have fun doing it. We'll spend the rest of this book discussing agile frameworks like Scrum and Kanban, but you need to know that they don't work unless the teams using them understand the agile principles upon which they're based.

Agile Principle #1: Focus on Creating Value for the Customer

What's it all for? If you're building a software product, the answer to that question is your customer.

That code you're working on every day? You're writing it to make someone else's life better, easier, happier, or safer. If your company makes you feel like you're only working to maximize profits, it's time to look for another job. If you don't consistently talk to customers, show them your work, get feedback, and iterate, then it's time to change your processes. **Good products solve customers' problems.**

Agile Principle #2: Incorporating Customer Feedback

I worked at a start-up called Opower in the mid-2010s that used behavioral science and the phenomenon of "social proof" to get people to save energy.[8] Social proof in this context was like good peer pressure. It's the idea that humans tend to copy their peers' actions. I was the Agile Program Manager for our main product, the Home Energy Report, which compared people's energy usage to their neighbors' consumption. If your neighbors were doing more to save the planet, you would receive tips to decrease your energy usage. These suggestions included simple things like turning on a fan instead of the air conditioner.

During my time at Opower, we sold the product to our first Asian customer, China Light & Power in Hong Kong. When we showed them the existing Home Energy Report, we got our first key piece of customer feedback: It needed to be translated from English.

That might seem obvious in retrospect, but Hong Kong had been a British colony and territory until 1997, so we assumed most of the inhabitants spoke English. Boy was that wrong! As of Hong Kong's 2016 census, only 4.3% of the population writes and speaks English. Thirty-eight percent have only some basic conversational level of fluency.[9]

Sending our reports in English would have immediately made them 60 percent less effective. That's some key customer feedback.

Simply asking customers what they want isn't enough. Most customers either have a grand vision that would be impossible to build or they don't know what's possible until they see it. **It's much easier to get valuable insight if you show customers an example of what you're doing.** Show them a design, a mockup, an early version of your work, or a prototype. Let them walk through your prototype themselves and talk through what's going on in their heads. Don't ask leading questions. Listen.

If you get comfortable sharing early work, you'll save yourself a lot of time, learn how to better solve your customers' problems, have a more valuable product to sell, and be more successful at the end of the day.

Get feedback from real customers *in your target audience* early and often. Doing so can mean the difference between releasing any old feature and releasing a feature customers love.

Are engineers actually required to talk to customers? While it's true that many don't, the best teams get their engineers involved with customers. Hearing how your work is directly impacting other people is incredibly valuable and rewarding. We'll talk more about how to get feedback in Chapter 3.

Agile Principle #3: Continuous Improvement

Continuous Improvement is my favorite agile principle, and here's why: On agile teams, we admit from the start that we're going to get some things wrong. That's okay, because we can and should iterate and evolve over time.

I find this idea incredibly refreshing. No matter where you start, you can make yourself, processes, culture, architecture, and everything else better over time, little by little.

You can't release a flawless product right away. The whole point of agility is to take what you've built, get feedback on it, and improve it. The more customers you talk to, the better your product will get over time.

Continuous improvement also applies to how you build your product. Many teams use a meeting called a "retrospective" to help examine their own practices.[10] In retrospectives, we ask ourselves:

1. What's going well?
2. What's not going well?

3. What actions can we take to improve the things that aren't working?
4. Which action item(s) will we focus on next?

These four questions, over time, can take a team from non-performing to high-performing. They can make a mediocre product highly profitable and take a workplace from having a culture that's "fine" to one where people love to come to work. Little by little, improvement happens, and over time, small changes equal big results.

Agile Principle #4: Iterative Development

Remember waterfall development, when we shipped huge changes after years of heads-down work? One big problem with this approach is that over the course of time, things change! Customers' needs change; the market changes as new competitors and features arrive; teams themselves change over the course of a long release; new members bring new ideas and product directions.

Today, high-performing teams work in much shorter release cycles (less than four weeks). Then, they iterate what they've built, adding valuable functionality at each iteration. This model allows customers to try out the features and react to them. Their feedback, in turn, provides insight into what should be changed or built next. It also provides clarity around what customers find valuable.

Setting up that short-cycle, iterative feedback loop is important. Get your product in front of customers at each iteration. **Make changes, get feedback, and repeat.**

Agile Principle #5: Responding to Change

Perfectly stable teams, stable markets, and projects that follow the project plan exactly don't exist. I've always loved the Mike Tyson quote about planning, "Everyone has a plan until they get punched in the mouth." As an engineer, you'll deal with change on a daily basis. Some changes are big ("we won't be working on this product anymore"), some changes are smaller ("this library must be updated because we've found a security issue"), and some changes feel like getting punched in the mouth. The only constant is change.

So, the name of the game is not "how do we minimize change?" but rather, "How do we best respond to change?" **The better we get at responding to change, the easier each change becomes.** The frameworks we discuss later in this book help us respond to that inevitable change.

Teams must be focused and empowered to become high-performing and effective. Being empowered means making decisions about HOW and on WHAT teams work.

HOW: If a team wants to meet at a coffee shop one day each week, they do that. If they want to use Scrum instead of Kanban, that's fine. If they want to release every two weeks and hold a happy hour every other Friday to celebrate, who are we to judge? Over time, these decisions add up and can lead to a much happier, more productive team.

WHAT: Some teams are handed requirements and told to go work on them, but on empowered teams, the team figures out what to build. An agile Scrum team has very few special positions, but the Product Owner is one of them.[11] Product Owners are sometimes called the "CEO" of their product. They're in touch with customers and business stakeholders, and they own the backlog—what's in it and its priority. Product Owners address questions that crop up from the team; no complicated chain of command is in place to make decisions. They also lead the team in doing product discovery work, which we'll dive into in Chapter 3.

Teams should also be dedicated. The more projects or work in progress a team has, the longer it takes to get everything done. Let's look at a simple non-software example. Imagine a busy highway system in Los Angeles or Washington, DC. On weekends, fewer cars drive on the roads. People can drive the speed limit and have space to pass slower traffic. During rush hour, however, more cars on the road mean everything grinds to a halt. When I lived in the Washington, DC area, I could get downtown from Springfield, VA in twenty-five minutes on the weekends. During rush hour, though, that same route would take me over an hour. Obviously, the amount of traffic clogging up a road changes the flow of traffic.

This concept can also be applied to work. The more projects a team has going at one time, the tighter the schedule and the less room they have for error. More work in progress means all work moves slower. We'll prove this concept mathematically in Chapter 6.

The combination of both empowered and dedicated teams is powerful. Organizations that build empowered, dedicated teams are rewarded with productivity.

Most of these principles are great for personal efficiency too. If you want to get more done, focus on one task at a time before moving to the next one!

Agile Principle #7: Simplicity

Agile software development maximizes the following (not an exhaustive list):

- The number of meetings we're NOT in
- The number of lines of code we DON'T write
- The pages of documentation we DON'T produce
- The number of dependencies we DON'T have
- Time spent on the most important tasks
- Team communication effectiveness
- The speed with which we remove team blockers
- Each team's decision-making abilities and empowerment

With agile development, simplicity guides everything we do as a software development team.

"The simplest solution is almost always the best solution." - Occam's razor[12]

Agile teams strive to incorporate simplicity in everything they build. **The smaller and more decoupled (free from dependencies on other system parts) a solution is, the easier it is to add features.**

Agile Principle #8: Eliminating Waste/Lightweight Processes

Have you ever heard of essentialism? Essentialism is like minimalism for your time. Whereas minimalism eliminates the extra "stuff" in your life, essentialism eliminates the extra activities in your life.[13] It's the idea that you keep doing only things that bring you joy or that move you toward your life goals. You can always make more money, but you can never get back your time. So, how do you ensure you're using the time you do have wisely? This principle of eliminating waste, applied to software development, is essentialism.

What does this concept look like in reality? Waste elimination can happen at any and all levels of your work. For instance, take a look at how your team is building their software. Do they go to extra meetings that don't contribute to the project? Could they be automating test cases so your testers can focus on other things? Do some steps or extra clicks in the product waste time? Do some people involved in your processes not add value? If you answered yes to any of these questions, eliminate those things. They're all waste.

It's very important to re-examine your processes, teams, and tools often. A heavyweight process or product isn't created overnight. It happens little by little, piece by piece as you scale. **Agile teams strive to have "just enough" process and err on the side of having too little.** In that way, you save yourself time: your most important non-renewable resource.

Agile Principle #9: Technical Excellence

Technical debt ("tech debt" for short) is what happens when teams decide to cut corners to ship faster.[14] Tech debt builds up when teams skip test case automation or don't upgrade libraries. It also builds up when teams choose a suboptimal architecture in order to release code faster. Customers can't see tech debt, so it's easy to forget, and it may not get fixed until the software starts to break.

Over time, tech debt delays development and makes the system less performant. For example, teams may start a new feature only to realize the existing codebase doesn't include automated test cases, so, they have to spend time writing them. The feature they are currently working on will therefore take longer to release.

Or, they cut corners on architecture, and as the number of customers grows, the system can't handle it. Sometimes, tech debt happens even when engineers do everything right. The bigger the system gets, the more the code must be refactored. Over time, tech debt builds up and causes customer problems. Eventually, the only way to get rid of it is to embark on huge refactoring or re-architecture projects, which take place instead of new feature development. Tech debt can't be entirely avoided, but it can be reduced.

"Technical excellence" describes the practices engineers use to avoid tech debt. Some of these practices are:

- Writing automated test cases that cover each feature, including performance testing.
- Documenting the code.
- Updating libraries on a regular basis.
- Writing code that doesn't have a lot of dependencies on other system segments.

- Building the infrastructure needed to automate code deployment.

Incurring tech debt is easy, but getting rid of it is a real pain in the a**.

Agile Principle #10: Sustainable Development

Sustainable development means working at a steady pace that won't lead to burn-out.

I'd like to address a bit of contradictory language used in agile software development now. Scrum has a concept called a "sprint,"[15] which is a time box in which teams pull in a reasonable amount of work and then commit to getting it done. Sprints are generally one to four weeks long, and many teams today run two-week sprints. The word "sprint" implies that teams are working as fast as they can to reach the finish line, but that's not exactly the case. Teams are working a fast-but-sustainable pace—one they could continue forever, if needed.

Let me diverge a bit and give you a small piece of related advice.

As humans, we're not built to work as much as we do, especially those of us in the tech industry. When I worked at IBM, engineers commonly bragged about working every evening and on weekends. This mindset created a culture in which people felt they were falling behind if they didn't work insane hours, so everyone did. We'd get burned out and sacrifice time spent with our families, pursuing hobbies, and volunteering.

When people take time to relax, they're far more productive than if they fill every hour of the day with work. Full stop. You're not doing yourself or your company any favors when you work overtime. In fact, studies have shown that knowledge workers can focus and be productive for about four to six hours per day;[16] that's it! After that, we tend to be far less effective and productive without rest.[17]

You may not be able to work only four-hour days, but you can carve out your nights and weekends to relax. Individuals working sustainably influence teams to work at a sustainable pace too.

Agile Techniques

When some people use the term "agile," they're referring to one or more agile frameworks developed to help teams utilize the agile principles. Dozens of frameworks fall under this "agile umbrella." A few of the most popular ones are Scrum, Kanban, Scaled Agile and eXtreme Programming (XP).[18] Certain capabilities also belong under this agile umbrella. "Business Agility," for example, describes a scenario in which an entire organization (marketing, finance, sales, software development, etc.) embraces agile principles and culture.[19]

So, when you hear someone say, "I practice agile software development" you can ask, "What framework do you use?" or, "How do your teams embody the agile principles?"

The Agile Umbrella

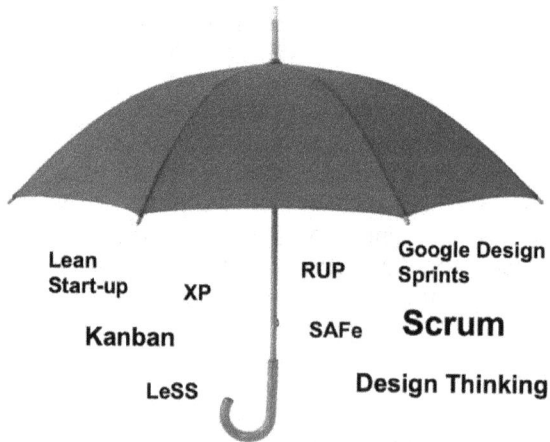

Lean Start-up   XP   RUP   Google Design Sprints

Kanban   SAFe   **Scrum**

LeSS   Design Thinking

Agile Overview: Reality

In reality, teams tend to forget about these important agile principles. They adopt a framework like Scrum or Kanban and then use it without thinking about the principles it enables. These aren't magic frameworks; in truth, the framework structure you use doesn't matter as much as you think! If your team embodies the above principles, you'll have a successful agile team. These frameworks simply make teams that understand the principles more efficient.

Agile Overview: Survival Tips

1. Remember the principles - For the rest of this book, we dive into specific frameworks and steps you can take to become a happy, successful engineer. But remember, none of these frameworks are magic. **Without an understanding of these underlying agile principles, they won't be effective.**

2. Frameworks - When starting a new framework, adopt it as-is for a few months. See what the framework has to offer, and then adjust it as necessary from there. Too often, teams half-adopt a framework. Then, they decide it doesn't work and use that as an excuse never to try another agile framework again. In this case, the framework doesn't stink; the adoption of it does.

3. Evolve as you go - Always continuously improve your processes, which means you may make a departure from a framework's suggestions. Doing so is healthy and almost inevitable. Some changes will propel you forward, whereas others may not. Nix the bad changes and keep honing the others. This strategy works well on agile teams and for personal development too!

4. Transformations - If your company is adopting an agile framework for the first time, the following are important:

   a. Ask your manager if you have buy-in from all levels to iterate released products. If not, this transformation may be difficult, so brace yourself.

   b. Set up processes to talk to customers often. No matter how large your company is, hearing directly from your customers is always important.

   c. Make sure your team understands the agile principles before you adopt an agile framework.

   d. Does the organization already enjoy a high level of trust? If so, it will make things easier. If not, do you really want to work there and not be trusted and empowered to do a good job?

Chapter 2: Agile Teams

Teams that are focused and empowered to make decisions are high-performing and effective. Engineers searching for their next position can keep an eye out for these teams, and entrepreneurs can strive to build them. Let's talk about a few things to watch for.

Agile Teams: Key Concepts

High-performing and happy teams tend to be dedicated, self-organizing, empowered, and cross-functional.

Dedicated teams work on one project at a time. They move far more quickly than teams working on more than one project. To learn why, visit the principles section in Chapter 6: "Kanban," which discusses Little's Law. Little's Law mathematically proves that teams get more work done when focused on fewer tasks.

Empowered and self-organizing teams have everyone they need on the team in order to make decisions. They determine how to get their tasks done based on what's right for them. The team determines what to build, and no two teams are alike.

Cross-functional teams have the skills they need to complete their projects. The developers have the skills to do the work (or can learn them), the team has testing skills, and a Product Owner understands the product's business case. A Scrum Master or project manager will have the organizational skills the team needs, and team members may also have design or user experience skills. Strong teams also have go-to-market capabilities (sales, marketing) embedded in their teams.

Great team members can take on more than one role in order to move the work along, an ability called *polyskilling*. For example, developers commonly act as testers. Or, UX designers may have some front-end development skills they can use to build web pages. A strong cross-functional team has backups for each skill set, and they also have a learning mindset that allows people to jump in where needed. After all, we all take vacations or switch jobs now and again!

In general, agile teams consist of no more than ten people.[1] Smaller teams suffer from knowledge silos, which is when they don't have enough people to back each other up. On the flip side, for larger teams, communication becomes difficult. However, larger teams are okay if they remove dependencies on other teams. Dependencies slow teams down and create uncertainty. When in doubt, optimize for removing dependencies. If needed, teams can grow to not more than fifteen people;[2] after that, communication becomes really difficult.

Forming Agile Teams

In most organizations, managers form the teams, and members are placed onto the team that best matches their skillset. If you're lucky, your team will be stable for long periods of time and you'll like them. Most of the time, engineers don't have full jurisdiction over with whom and on what they work.

If you're a new engineer, you may not be in a position to change the way your organization forms teams. However, understanding your options may help you influence future decisions. If you're an entrepreneur or are in a position where you have influence over team formation, this section is for you. Let's look at a far better way to form agile teams. You *can* let your people choose with whom and on what they work.

Self-Selection

Wait, what? Let people choose? That'll never work! How will we make sure the teams have balanced skills? How will we guarantee all the work (including the things no one wants to do) gets done? How do we make sure people don't feel left out? A million things can go wrong. It's doomed to fail . . . right?

It *does* work. Using a process called self-selection, you can form balanced, high-performing teams over the course of an afternoon or a day. This is the process we use to form teams in my class.

What is self-selection?

Self-selection is an iterative process that really does let people choose with whom and on what they work. People are given team options, and then they decide for themselves which team to join. I've run dozens of these self-selection workshops, and they work. Every. Time. Self-selection works for small agile organizations, Fortune 500 consulting firms, and academic classrooms. Because you're asking people to make their own decisions, they end up matched up with work they love (or at least like). So, morale goes up and work moves faster.

The Self-Selection Process

The self-selection process culminates in a self-selection event. I recommend reading David Mole and Sandy Mamoli's book *Creating Great Teams: How Self-Selection Lets People Excel*.[3] In it, they describe the process from start to finish. In the planning period, the number of teams and their responsibilities are decided and a team leader is chosen—usually the team's Product Owner. Having the Product Owner be the team leader makes logical sense because the Product Owner has the depth of market, technical, and stakeholder knowledge to excel with that team.

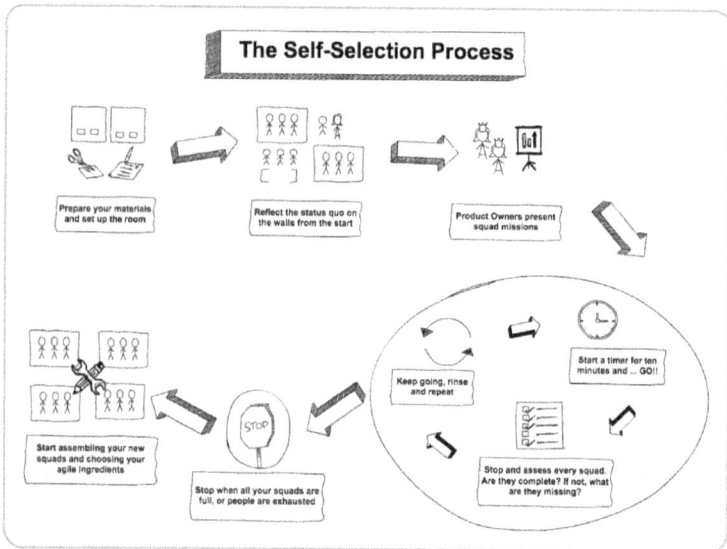

Image Courtesy of Nomad8

Product Owners then create a team one-pager, which describes the current project or product and the skills needed on that team. The paper might include a few fun facts about how they envision the team working together or why members would want to work with them. Individuals choosing teams also come to the event with a one-pager about themselves. They'll list skills, interests, and fun facts too. Creativity is encouraged at this stage! Your organization can put together templates for both one-pagers, if that's helpful. If you do, add a few fun categories like "What's your favorite movie?" or, "What would your Patronus be and why?" During selection rounds, people can read their teammates' one-pagers at their leisure and may just learn something new about a close co-worker (your favorite movie is also *The Princess Bride*?)!

The hardest part of preparation is convincing people to try self-selection because it's so different from what they're used to. Even the people who get to pick their teams are nervous about it. I've had more than one person privately ask me, "Can I just give you my top three choices and have you pick?" However, everyone was happy I didn't do that after they'd gone through the self-selection process.

Self-selection doesn't work if the executive team in charge doesn't buy in. If managers insist on being at the event or threaten to move people around after the event, they aren't ready to try it. Trust and empowerment are core values for agile teams and self-selection, so management needs to be committed to standing by what their people choose. Managers who can get over feeling like they're losing control will reap an abundance of rewards throughout this process.

Self-Selection Event

The self-selection event takes from a couple of hours to up to a day, depending on department size. A team of 200 engineers might have an event that lasts all day, whereas a team of forty engineers will likely be done after two or three hours. On the day of the event, the agenda is:

1. 15 Mins | Gather and explain logistics.
2. 45-60 Mins | Team leads pitch their teams and answer questions.
3. 10 Mins | Self Selection: Round 1
    1. People place their individual info cards near team info cards.
    2. If the organization is distributed, location captains digitize choices in real-time.
4. 10 Mins | Round 1 Assessment
    1. Are any teams complete?
    2. Point out gaps.
5. 10 Mins | Round 2
6. Repeat! Stop when teams are full, nothing is changing, or you run out of time.

Team selection is typically finished after three rounds.

Round 1: Everyone chooses their first choice, and then you marvel at how uneven the teams are.

Round 2: Stalemate time! No one moves because they're waiting for other people to move first. Moderators step in and help determine why people love their teams. They find out what levers can be pulled to help make people happy with their second or third choices. Can you move an area of work to another team? Could several people move at once, so friends can work together? Can someone learn a new skill? All these problems get worked out during Rounds 2 and 3.

Round 3: Change happens. People start to move to their second or third choices. Deals are struck. We usually end this round with everyone happy about where they landed and a few open gaps. Those gaps are natural and can be filled with open hiring positions.

Self-Selection Rules

One rule is essential to make this process work. Introduce the rule early and often. Remind everyone of it, and then post it in the room during the self-selection event. **That rule is: "Do what's best for the company."**

I have seen people spend two rounds on a team they'd really like to be on and then walk back to the team they came from because they had a specialized skill set. Astonishingly, these people are still happy with the self-selection results. Why? I suspect it's because they had the chance to make where they wanted to be known. Then, they very publicly "took one for the team." They're a hero, and that feels good too. Sometimes, self-selection can be the beginning of a very fruitful development conversation. Now, your manager knows which team you want to be on and can help get you there.

Other rules that make sense for self-selection might be rules around team size or skills. For example, "Each team must have at least one tester," or, "Teams must consist of at least five people."

The only mandatory rule is the first one: "Do what's best for the company."

Self-Selection Results

As with any change within an organization, it is important to measure before and after the event. Take a measurement of throughput and team happiness before the self-selection event; then, measure the same metrics afterward and compare. I send a survey out after every self-selection event with a couple of key questions.

Are you happy with your team? Ninety-five to 100 percent of people say they're happy with the teams on which they landed. The other respondents are usually okay with the process—neither happy nor unhappy. Occasionally, you'll have a participant who's dissatisfied with the results. The reason is usually very individual to that person and should be addressed with them. Making everyone happy with any given change is difficult, but the number of people who walk away satisfied after self-selection is worth it.

What primarily drove your self-selection choices? We had thought most people would choose teams based on the project type that appealed to them, but instead we found that most people made their decision based on the company's best interests as well as who would work on the team with them.

What Primarily Drove Your Self-Selection Choices?

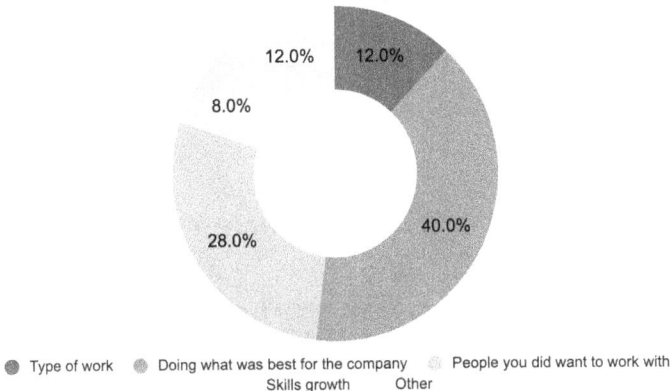

Legend: Type of work • Doing what was best for the company • People you did want to work with • Skills growth • Other

Opower's Post-Self-Selection Data - What primarily drove your self-selection choices?

Skeptics are afraid that people will work on what they want and leave the company high and dry, but as you can see here, 40% of the respondents primarily chose their team based on the company's best interests. Secondarily, they were driven by the people who would work on the team with them (28%).

Self-Selection Fears

People have a variety of fears regarding self-selection. If you're planning to run an event, head over to my blog (amberrfield.com) to learn more about how to address those fears. For now, I'd like to leave you with this.

A Note on Self-Selection and Happiness

It's important to note here the positive impact that empowerment can have on our happiness. I'd like to share one quote from an engineer I worked with at Opower, who was maybe our biggest skeptic regarding the process. Every day, he would come in and have a new question or try to poke a new hole in what we were doing. After going through the process once, here's what he wrote in his feedback survey:

> *Looking at the outcome of the process, it is difficult to argue against–each team appears well-suited to meet its short-term needs and the teams are reasonably well-distributed. If we had done this through another process 100 times, I believe that most of the outcomes would not have been as good as this one.— Josh E*

Agile Teams: Reality

Most organizations do a good job of creating decent agile team structures. However, they may not understand how to fully empower their teams. Maybe managers or executives still make a lot of the decisions and Product Owners aren't empowered enough to make quick decisions. Traditionally, managers were the decision-makers in an organization, and it's hard for some to give up that control.

Some of the people on a team might be shared, especially in smaller organizations. This structure can slow work down if you have to wait for their attention before making progress on any given project. User experience (UX), for example, may have a centralized team that serves several teams. DevOps or release engineering teams (see Chapter 7 for more information) might too. If shared teams have a transparent backlog and a path to discuss priority changes if needed, that's okay.

Many organizations claim to be "agile," but they don't embody and lean on the principles we talked about in Chapter 1. For example, project managers become Scrum Masters when they first adopt agile practices.

Scrum Masters and project managers have similar roles, but very different attitudes. Project managers keep the teams on track and assign work to team members, whereas Scrum Masters are servant leaders who empower their teams and unblock them. Scrum Masters help teams take advantage of the agile frameworks and principles. If the wrong person is in the role, the team won't be high-performing.

Very few companies have tried self-selection. As I mentioned, the toughest part about the process is getting people to try it. Once they have, most are sold on its benefits and would do it again. The process itself, however, can get messy in the middle. The event can feel chaotic at times, but that chaos is worth the end result. It's important to set expectations that self-selection may feel a bit uncomfortable. It's also important to have great mediators in the room who can help with stalemates. Some teams have reported to me that self-selection didn't work for them, but when we dig in, the problems almost always boil down to lack of trust or preparation within the organization, not the self-selection process itself.

Agile Teams: Survival Tips

1. Job Searching - It's easier to find an organization that has great, empowered teams than it is to create one. What should you look for when searching for a new job or team?

    a. Cultural fit - Do you like the people you're talking to on the team? Do you see signs that you'll like the work environment? I interviewed for an internship position at Microsoft when I was in college, and one of the interviewers mumbled throughout the interview. Honestly, he seemed annoyed to be talking to me. Needless to say, I didn't pursue that job. In general, how I'm treated during job interviews has guided me to fun places where I loved working.

    b. Managers who trust their people - Ask your potential teammates whether their managers trust their opinions. Ask them to talk you through how the team works and how they solve conflicts. If people outside the team make a lot of what should be the team's decisions, or the team often gets blocked, those may be bad signs.

   c.   Happy engineers who don't work all the time - Do your potential teammates seem genuinely happy? Are they enthusiastic about what they're doing? Do they have any complaints about the organization, the team, or their managers? Are they working night and day, or do they have time to breathe? If people are bragging about how many hours they work, think twice about working there. Dedicating every waking moment to a project can be fun, but everyone burns out sooner or later.

   d.   Dedicated teams - Ask flat-out whether your new team is dedicated to one project. Focus is important.

   e.   Don't settle - If you're not excited about the job and the team, keep looking. Your skills as a software developer are valuable. Plenty of teams that would love to work with you are out there. Wait until you find a team to which you can say "hell yes!"

2.   Your Current Team - Let's say you just started on a team and it's not perfect (none of them are). What can you do to encourage healthy team dynamics?

   a.   Focus - Focus on one task at a time, get it done, and then move on to the next one. Your team should do the same thing with projects.

   b.   Sharing - If you're shared across more than one team, over-communicate with each team. Better yet, set up a backlog board showing everything you're working on and keep it up to date. That way, everyone knows where they are on the priority queue.

   c.   Kanban - Kanban (discussed in Chapter 6) is a lightweight way to adopt agile practices. See if your team is game to start using Kanban. At the very least, you can use it for your own to-do list and limit how much you're working on at one time. I use Kanban for my personal to-do list and it works great!

d. Block Time - Block time on your calendar so you can focus on deep coding work.

e. Technical Debt - It's a good rule of thumb for teams to always spend at least 20 to 25 percent of their time on tech debt.[4] Addressing a little bit of tech debt at a time, alongside feature work, is much easier than trying to do it all at once. Be disciplined and work on technical debt along the way. Your future self will thank you.

3. Be the Change - What if you feel passionate about agile teams and culture, but your company hasn't adopted them yet? Can you do anything? Yes!

a. Talk to your manager and team leads – Sometimes, all it takes is one employee who feels passionate enough about new ideas to drive them forward and gain support.

b. Work with your organization's influencers - Get them on board. Management may push back on certain concepts, but come up with plans to get them to admit that changes are safe enough to try.

c. Add credibility - Find credible educators who can explain agile benefits to the management team. At National Geographic, I brought in Sanjiv Augustine from Lithespeed. He had an excellent pitch on why agility was great for the executive team. He got them to buy in, which made the agile transformation much easier for me.

d. Run a pilot - Make a change on one team versus the entire organization. Prove that it works, and then expand to other teams.

e. Still not getting traction? - If you hit a wall, consider moving to a more open-minded organization. A lot of places would love to hire a forward-thinking software developer like you, so find a company whose purpose you're passionate about and negotiate a raise in the process. That's what you call a win-win.

4. Be Prepared - If you're participating in a self-selection event, keep an open mind and come prepared. Have two or three teams in mind on which you'd be happy to end up. Think about what skills you can contribute or that you might want to develop, and ask questions before and during the process. Think about who you'd like to work with and why. Come with options. It's your choice, but the process will be easier if you know you have many great possibilities.

5. Advocate for Yourself - Don't be shy about lobbying for yourself during self-selection. Actually, this advice doesn't only apply to self-selection; it's important throughout your career. No one will know what you want until you tell them, and the worst answer you can get is "no" (which usually means "not yet"). When you ask for what you want, it can start a chain reaction that leads you to that place. Maybe you won't get that raise or promotion right away, but now your manager knows you want it. Now, she can advocate for you when future opportunities come up. I'm speaking especially to females in tech, as we tend to be less outspoken about what we want. Please speak up! As a manager myself, I find it so much easier to help you if you help me know where you'd like to be.

6. Talk to Everyone - If you're lucky enough to run a self-selection event, spend a lot of time talking to people beforehand because it may take months to make everyone comfortable with the idea. Talk to them in groups and one-on-one. Keep a "frequently asked questions" (FAQ) document with answers to all the questions you receive. Everyone is uncomfortable with the process before they try it. The only way to make people more comfortable is to tease out their questions and fears so you can answer them.

Chapter 3: Finding Product/Market Fit

Everything we talk about in this book—absolutely everything—is meaningless long-term unless you get the answer to one question right. That question is: "What should we build?" The quest to answer this question is called "product discovery." Discovery can be a phase at the beginning of a project, or you can infuse it into the very fabric of everything you do.

You build products to solve a problem for your customers. To do this, you must:

1. Understand who your customers are and what problems they need to solve.
2. Build an easy-to-use product that solves a problem for those customers.
3. Price the product affordably for those customers.

Customer Problem + Effective Solution + Value = Successful Product

This convergence of a valuable solution with customers who are willing to pay for it is called *product-market fit*.[1]

Engineers at a start-up are obsessed with finding product-market fit. Engineers at a larger, more established company may work on a product that has already found product-market fit. The techniques in this chapter can be used at any stage to figure out what to build, even if you're building a new feature for an established product. I encourage new engineers working in any capacity to incorporate some of these ideas.

Discovery: Key Concepts

Let's first have a little fun learning how *not* to do product discovery. The chances that the first company you work for will do one or more of the following while figuring out what to build are decent:

1. Guess - You may be very intelligent, but chances are you don't represent most of your target market.

2. Executive decree - Especially true if the idea was generated after reading one article or sitting on the plane next to "a guy."

3. Sales or marketing input alone - Ideas come from potential customers at a fast and furious pace. Building ideas without actual financial commitments behind them can waste precious time.

4. Engineer's choice - Building a feature without customer or user experience input will result in a product that may only work for a small number of users.

5. Build your first idea - Ideas get better with time and iteration.

6. Build what the first customer tells you they want - Hearing an idea once is interesting. Hearing it twenty times makes it a trend upon which you can act. Soon, you'll be introduced to a key concept that's worth pondering: Customers don't know what they want until they see it!

7. Build what your mom or friends want - Your mom will always tell you your work is brilliant, even if it isn't.

8. Dart board - This is not a great way to decide what to build, but it's not a horrible way of coming up with estimates on the spot. Just kidding. (But I have used a dart board to make an estimate before.)

While these seem like quick and easy ways to get a project off the ground, chances are good that the product or feature won't be successful. So, how do you run a discovery phase and maximize your chances for success?

Believe it or not, you're already familiar with how to do product discovery. In fact, you've already done it! Remember the third-grade science fair project you did (with quite a bit of help from your parents)? That project taught you how to use the scientific process.[2]

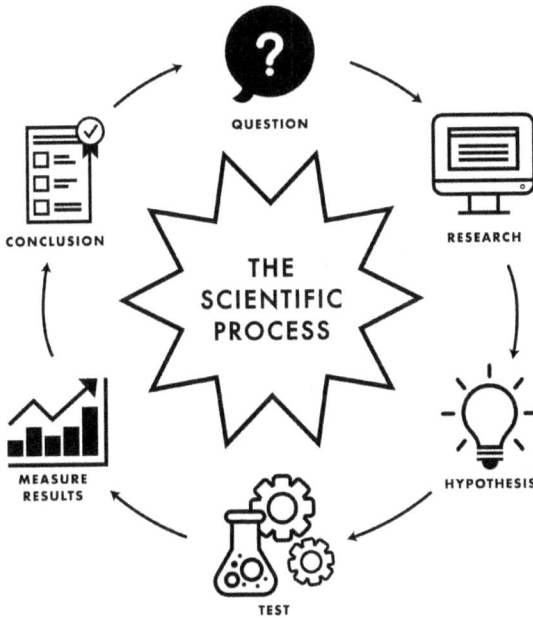

THE SCIENTIFIC PROCESS

QUESTION — RESEARCH — HYPOTHESIS — TEST — MEASURE RESULTS — CONCLUSION

Surprise! The scientific process works for developing vaccines *and* product discovery! It has the same basic steps as many of the product discovery frameworks, but to build great products, we use the scientific process with an extra dose of customer empathy.

An Engineer's Role in Product Discovery

In many companies, engineers are insulated from customer conversations. They hear about customer requests from a Product Owner, the sales team, or someone else, and then they design their products based on that secondhand knowledge. However, a lot can get lost in translation when engineers aren't listening to customers. The more you hear directly from customers, the better you can design your systems. Product Owners and UX will often take the lead in customer conversations, but engineers should be in the room listening, taking notes, and asking questions. As an engineer, even if you stay silent during the entire interview, you'll learn a great deal from your customers.

You'll also start to learn how your products are impacting your customers in a positive way, and that's the best part! Nothing is better than hearing about how your work is improving your customers' lives—

that feedback can give you an emotional boost that lasts a long time. **If you take only one thing away from this book, I want it to be this: Talk to your customers.**

Who Is Your Customer?

Often, when you think of your customers, you probably picture the end-user. If we're talking about Twitter, it's the person tweeting or reading tweets. If you work for Amazon, it's the shopper. That's all you need to know, right? Wrong!

To build great software, you need to get a lot more specific about who your customers are. You must understand what they do, what they like, and where they live. You must also consider other types of customers: vendors, partners, and internal customers.

End-Users

End-users are the people who use the product on a regular basis; they're the ones who log in and gain some kind of value from the system. Always test and get feedback from end-users, even if they aren't the ones directly paying you (more on that next).

How do you determine who your end-users are? First, make an educated guess about who your target customer group is. Then, talk to a few people in that group, and hone in on their traits and characteristics. Keep in mind that your initial target group guess may change as you talk to customers.

Let's say you're building a financial app to teach kids how to handle their money. Who are your end-users? If you said "kids," you're both right and wrong. What ages should you target? After all, preschoolers' and high school students' needs are very different. Also, where do these kids live? How much discretionary income do their parents have? Get as specific about your users as you can. Consider age, gender, demographics, income levels, location, what they do in their spare time, and what they care about.

Don't stop at "high school students," for example. "High school juniors and seniors from urban households living below the poverty line" is far more useful. You'd likely assume that high school students in general have access to and basically live on their phones every waking hour of the day, but that second, more specific group might not even have internet access. The more you can zero in on your customers' traits, the better your testing and product will be. You can always change your target audience as you collect more data to support that decision.

Vendors, Partners, and Buyers

Many times, companies don't sell products directly to end-users. Instead, they sell through partners, or they sell to vendors. For example, clothing manufacturers often don't sell clothes to customers; they sell to a store (e.g., Target) that sells to the end customer. If your business is selling to an intermediary (a vendor), you need to impress two people: your end-user and your vendor.

Also, in large companies, the person who buys your product (procurement) may not be the person who uses the product. So, you have to find ways to target the procurement department as well as keep your end-users happy.

Partners are similar, but they work with you to sell your product to an end-user. Singlewire Software (my company), creates an emergency mass notification platform that works with Cisco phones. We have a wide network of Cisco partners who work to sell Cisco phones as a package with our products. Therefore, we've got to keep Cisco and the schools, hospitals, and companies who use our system happy and educated.

Internal Customers

At IBM, my job was to customize a tool called Rational ClearQuest for IBM's Systems & Technology Group. My customers were other IBM employees; in other words, they were internal customers. Any group that delivers work to other teams has internal customers, and you can use the techniques in this section to find product-market fit. However, keep in mind that your end-users are your co-workers, so their happiness and your company's continued funding are the validation you seek.

Now that we understand who our customers are, let's discuss how product discovery works. We'll start with two very useful frameworks: lean start-up and design thinking.

Lean Start-up

Eric Ries, author of *The Lean Start-up: How Today's Entrepreneurs Use Continuous Innovation to Create Radically Successful Businesses,*[3] popularized the lean start-up framework in 2011. Ries based his book on the academic teachings of Stanford University professor and entrepreneur Steve Blank.[4] Lean startup helps entrepreneurs find a business model and product-market fit. "Lean" is another agile framework that focuses on eliminating waste and delivering only what the product needs. The manufacturing industry used it first, but today, it has been applied to many industries, including software product development.[5]

The lean start-up can be boiled down to three actions: build, measure, and learn. In other words, build something quickly, test it with customers, measure the results, and then learn from those results. Repeat.[6]

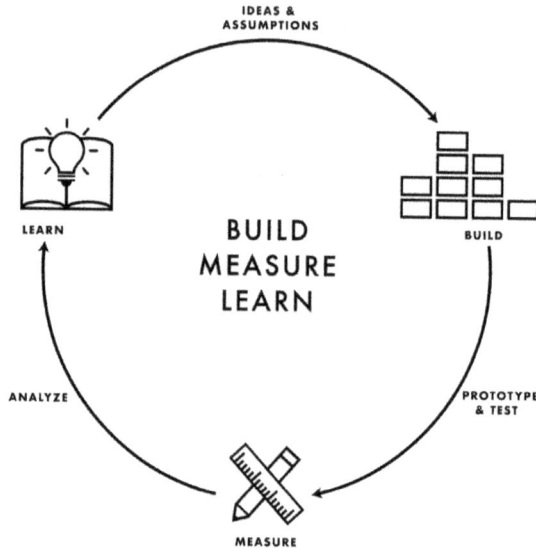

Start with a product idea. For lean startup, someone has generally come up with a product idea, but not always. If your team doesn't have a product idea yet, then skip down to the design thinking section. Design thinking helps you find customer problems to solve!

Step #1: Generate Assumptions

Next, write down the assumptions you have about your product.[7] An assumption is something you think is true but haven't verified. Make a huge list!

In one of my classes, we worked on a project for Capital One in which the idea was to build an app or web page to teach teenagers about financial literacy. A few early assumptions for that project might have been:

- *"I assume teenagers want to learn about financial literacy."*
- *"I assume most teenagers don't know the basics of personal finance."*
- *"I assume the best way to teach teenagers about money is to build an app for their phones."*
- *"I assume teenagers would visit an app on their phones more than once if it gave them coaching tips on how to save money."*

The key to successfully vetting your idea is listing as many assumptions as you can. When you brainstorm, write them all down, and don't judge your assumptions. What seems obvious to you may not be evident to your target audience. Think about what your customers need to do. What do they fear? What do they want? What are your product's benefits? What specific features do your customers need? What will the user experience be like?[8] Answering them will help you flesh out your own assumptions.

Step #2: Find Your Riskiest Assumption

The *riskiest assumption* is the one that, if false, would kill your entire product or feature and render any work you've already done useless. Which assumption, if proven wrong, would mean you'd have to build something else? That's the assumption you need to investigate right now.

In the Capital One project example above, the riskiest assumption is:

"I assume teenagers want to learn about financial literacy."

If that assumption is false, then nothing you build (app, website, semester-long course, etc.) is going to be successful.

Zappos.com founder Nick Swinmurn's riskiest assumption was that people would buy shoes online.[9]

Airbnb, the two-sided marketplace that matches travelers who need hotels with landlords who have extra space had two initial riskiest assumptions, one for each side of their business model:

1. People would be willing to pay money to stay in a stranger's house.
2. Landlords would be willing to let strangers stay in their homes.

In 2007, Brian Chesky and Joe Gebbia were grappling with those two assumptions. How would they find out if their assumptions were true or false?[10] They had to test them.

Step #3: Develop a Test

Once you know your riskiest assumption, you can develop a test for it.[11]

A good test:
- Tests your riskiest assumption
- Is developed and run in a matter of hours or days
- Involves real users

Right now, these criteria may seem impossible. Chances are you're currently wondering how the heck you put a working feature in front of your customers within a week. Simple. You *don't* do that. You can develop a test rapidly without writing a single line of code in many ways.

There Are Many Ways to Test

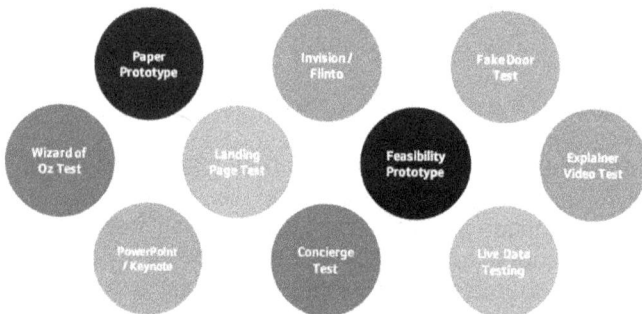

Image Courtesy of Charles Moore (Thrive Street Advisors), Capital One Labs

Prototyping

Many tests start with a prototype. Why? Because the best way to get feedback on a product, feature, or idea is to show it to a target customer for feedback.

No one knows what they want until they see it!

Put a prototype in front of your customers. Ask them to react to it, walk through it, and tell you what they think as they try to manipulate it. You can see what they see, note where they stumble, and hear their questions in real-time. Listen. That sort of test will give you data and insights you never expected. Without a prototype, customers will tell you they know exactly what they want. However, until they use it and pay money for it, you risk that their opinions won't lead to action.

So, let's talk about how to quickly build a prototype. This step is where your engineering skills shine in the product discovery phase!

Paper Prototype

One prototyping method that's quick, easy, and almost always available is a paper prototype.[12] If you're building a physical object, make it out of paper first. For instance, if you're creating an app, draw each screen on a separate piece of paper with a marker. Then, move the user through the various "screens" (i.e., pieces of paper) she would see given the actions she takes.[13] The only time you may not be able to paper prototype is if you're working with a remote team.

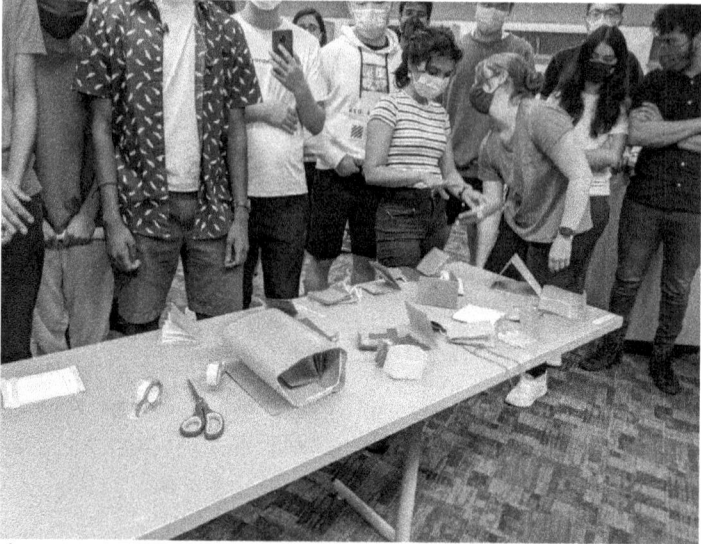

In class, we run the Stanford D-School's Wallet Project activity.[14] Students paper prototype "something meaningful" for a partner using iterative feedback sessions.

Paper prototypes are a great choice for very early testing, as you can build a paper prototype and test it with several customers in a matter of hours. The most important sign of start-up success is how many customers they talk to, and starting with a paper prototype can jumpstart those conversations.

Sometimes, people hesitate to put something imperfect into the world. Paper prototypes are never polished, but that's okay. When you show customers something that clearly needs work, you've set the expectations accordingly. In fact, this is a reason *not* to test with a polished prototype. If users see a pixel-perfect mockup, they don't know how or where to direct feedback. People are less timid about giving feedback when they know they aren't causing you extra work.

Lo-Fi Prototype

A step up from paper prototyping is the lo-fidelity (lo-fi) prototype.[15] Paper prototypes, sketches, and wireframes are a few of the many types of lo-fi prototyping. Use a program like Balsamiq, Figma, or Adobe XD to build wireframes (black and white drafts) or a prototype. These tools work great for software projects, and they can also be shared with remote customers via a Zoom session.

Be careful not to put too much time into this kind of prototype. Prototyping should be quick, so you can move on to the testing phase as rapidly as possible.

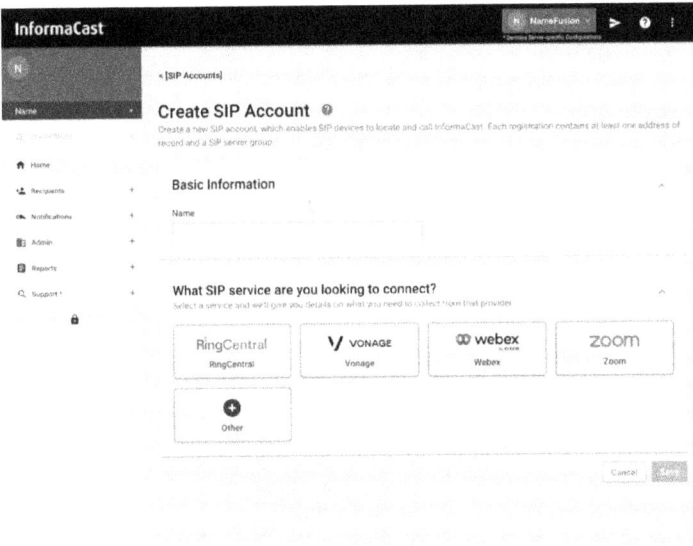

This black and white mockup was created by Singlewire Software's UX Lead, Ben Koca, for InformaCast's integration with various calling platforms including Zoom and WebEx. He made this in a matter of minutes using Adobe XD.

Landing Page Test

The landing page test allows us to gauge how many customers might pay for a product. First, you build a very simple webpage explaining the product. Then you funnel potential customers to a payment button and measure who clicks the button. If you can get enough customers to take action, you may have a viable idea!

The key here is to keep the page simple. Your call-to-action button may say, "Click here to subscribe," or "Sign up," or "Input payment details." Regardless, make it clear to users that they'll need to take a meaningful action, ideally involving money. When they click the button, you present them with a "thank you" message instead of a sign-up page. Let them know you're collecting data right now and will be happy to email them when the service is available. Make sure to collect their email addresses, as they may be excellent testers or customers later! Also, add an analytics tool like Google Analytics, which can show you where customers leave your page. If they don't subscribe, it may not necessarily be because they don't like your idea. It may be because they don't follow your pitch.

Next, set up some Google AdWords traffic to direct people to your landing page. The whole process can be completed in a couple of hours by someone with basic web development skills.[16]

Disclaimer: This strategy won't work in certain industries. For example, financial products are regulated—no one in the finance industry can sell a product that doesn't exist. These regulations are in place for good reasons! Be careful if you're in a regulated industry; this type of testing may not be available to you.

Concierge Test

Zappos founder, Nick Swinmurn, assumed customers would buy shoes online. (Remember we determined that was his riskiest assumption?) It seems obvious to us all now, but at the time, no one was buying shoes online. People thought an important part of buying shoes was getting to try them on and walk around in them. So, he came up with a novel way to test his riskiest assumption.

He went to a local shoe store and asked them if he could post pictures of their shoes online. If someone bought some shoes from his website, he would buy the shoes from the store and ship them to the customer. The shoe store agreed—what did they have to lose?

That's a pretty good test to determine whether people would buy shoes online! The website was easy to build: it simply required a minimal site, pictures of shoes, and checkout functionality. Swinmurn did all the real work manually.[17]

This is a concierge test.[18] To avoid wasting time, you carry out a service manually that will later be automated. Once you're sure enough people want your service, you can spend time building the full feature set.

Airbnb did something similar to test their two riskiest assumptions. An industrial design conference was taking place in San Francisco, and most of the local hotels were full. Brian Chesky and Joe Gebbia, two of Airbnb's founders, had some extra space, so they set up air mattresses and built a minimal website. Then, they offered conference-goers a unique "bed and breakfast" experience. The breakfast they offered conference-goers? Pop Tarts. This was the most minimal test they could come up with to assess their riskiest assumptions,[19] and it paid off. A strong culture of iteration propelled Airbnb to almost six billion revenue dollars in 2021.[20] Those six billion dollars started with a website, air mattresses, and a box of delicious frosted pastries.

Explainer Video

Sometimes, conveying your new product or feature without showing someone how it works from end to end is difficult, but that doesn't mean you need to build it.

When Dropbox founder Drew Houston tried to get venture funding, he had difficulty conveying his full vision to investors. Existing file-sharing solutions worked, but not as seamlessly as Houston envisioned. So, he mocked up a realistic-looking version of the product he wanted to develop and created an explainer video.[21] (Check out the video on *Tech Crunch* here: https://techcrunch.com/2011/10/19/dropbox-minimal-viable-product/.) That video showed his investors why his solution could be so powerful. His beta waiting list went from 5,000 people to 75,000 overnight.[22] Sometimes, a video is worth a thousand words (or, in this case, 70,000 potential customers)!

Step #4: Build Your Test (Quickly)

Using one of the techniques above (or any other idea), build your test and ready it for customers or customer interviews.

If you're interviewing, think about how you'll conduct your interview. Are you running it as the engineer, or will you have a UX designer or Product Owner take the lead? Where will your users be? If they're remote, you'll need a test that can be shared over a video call. What questions will you ask? What data do you want to collect? Will you record the interview? What technology will you need to have on hand?

Set up your test and the space in which you'll be testing. If this is your very first customer interview, practice with a friend. That way, you can work out any kinks in your testing process ahead of time.

You can do some great tests in the field. For example, you could go to a mall or a campus and ask questions of the people who pass by you. In this case, make sure your equipment is mobile and your test is easy to understand. If your customers aren't scheduled for a specific time slot, you may not be able to hold their attention for more than a few minutes.

Consider compensating customers for their time. If you spend an entire hour with them and you work for a company that has some money, a gift card may be appropriate. Such rewards can delight users and encourage them to come back for more interviews. Remuneration isn't necessary in many cases, but it's a nice touch. However, it may be necessary if you're having a hard time finding people to test. Or, if your target market is busy professionals, they may need to be compensated for their time.

Step #5: Run the Test and Measure Results

Now it's time to run your test. Some rules for a successful test are:

1. Run the test with customers from your target audience.
2. Run the test with as many customers as possible (double digits *at least*).
3. Choose metrics and measure results.

The first two steps are self-explanatory. You did some thinking already about your target customer, right? An important predictor of start-up success is the number of customers with which you test. This also applies to product and feature development. Try to reach double digits for every test. Use your network, go to the places where your target audience is, and plan ahead. For example, if your target audience is college freshmen on urban campuses, go stand outside a dorm. If your target audience is 31 to 45-year-old stay-at-home moms, go to the library during a tot program on a Tuesday. You can always find a creative way to connect with your target audience.

Measuring the Results (Innovation Accounting)

Before you begin your tests, develop a list of metrics that show the test is successful. For a landing page test, for example, you'd see how many people click your "subscribe" button after you make it clear they'll have to pay to subscribe. If you're testing a user interface change, you might measure how many people find a certain homepage feature without guidance.[23]

Vanity Metrics

Beware of vanity metrics. Vanity metrics are metrics that seem like a good idea to track but that don't give you actionable data. Examples of vanity metrics are number of downloads and total page views. These numbers measure your site traffic, but not your product's value. They don't measure whether someone will pay money for your product idea. Avoid vanity metrics at all costs.[24]

Step #6: Learn/Pivot or Persevere

This might be the most important step in the lean start-up process: Look at the test results to decide whether to pivot or persevere.[25]

Hold a meeting to decide what to do with the test results you've collected. Bring the entire project team, including a representative from the business, together to look at the data. If the data shows you're on the right track, that's great! In that case, you persevere. Return to your list of assumptions, find your next riskiest assumption, and repeat these steps.

Often, you'll find something you don't expect in the results. For instance, maybe your riskiest assumption is correct, but customers keep bringing up another problem that might be more interesting, painful, or valuable to fix. Or, you may find out your assumption was incorrect. The founders of Blue River Technology, Jorge Heraud and Lee Redden, originally wanted to build automated lawn mowers for golf courses, but when they talked to over a hundred customers in ten weeks, they realized golf courses didn't want their product. However, farmers desperately needed a way to automate killing weeds without chemicals, so Blue River pivoted and got three million dollars in venture funding less than a year later.[26]

If the data shows your assumption was false, you'll need to pivot. Your results will lead you to a new set of assumptions, so pick a new idea. Then, you can repeat the process over again to collect new data against your new assumptions.

Repeat!

With these six steps, lean start-up provides a way to iterate toward product-market fit.

Design Thinking

Design thinking is very similar to lean start-up. It came from IDEO, an innovative design company that has made improvements to many of the products you use daily,[27] and it has become a mainstream discovery method, particularly in the user experience (UX) and design community.

Lean start-up, as you'll recall, is focused on figuring out a viable business model. Design thinking, on the other hand, tries to figure out what **problem** you can solve for customers.

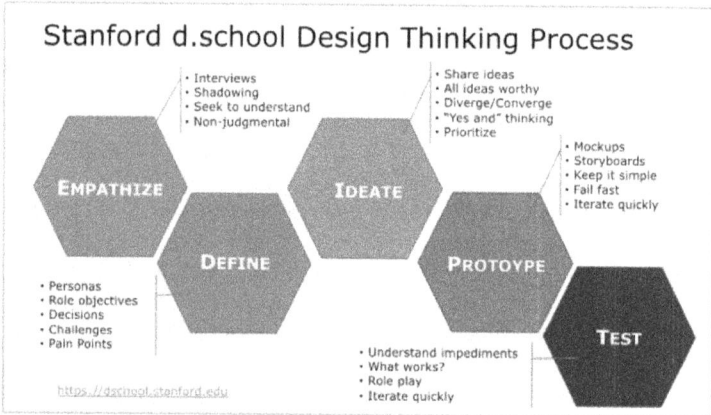

Stanford d.school streamlined design process (Legacy, circa 2012)

Note the fact that the last four hexagons above are Steps one through four in lean start-up. Design thinking adds a step at the beginning for gaining empathy for your customers.

The empathize phase is performed via interviews and shadowing. The aim is to gain as much insight into as many customers within your target group as you can. Only then can you form a set of assumptions and determine which of those assumptions to confirm. I find this concept of understanding your customers first to be invaluable; it's a great addition to the lean start-up process.

Design thinking includes advice for ideating, defining, prototyping, and testing. If you get lost at any step in the product discovery process, a great place to look for guidance is Stanford's *Design Thinking* blog[28] or IDEO's design thinking page. Airbnb, Uber Eats, and even the Oral B electric toothbrush were designed using design thinking.[29] We also used design thinking in Capital One's Innovation Lab to vet early-stage ideas.

Google Design Sprints

Another discovery framework I find useful is Google Design Sprints. Google Design Sprints come from Google Ventures. Jake Knapp, John Zeratsky, and Braden Kowitz discuss them in the book *Sprint: How to*

Solve Big Problems and Test New Ideas in Just Five Days.[30] Like design thinking, Google Design Sprints follow the same basic steps as lean start-up. However, each iteration can be done using a predictable pattern within five working days. Here's how the week is laid out:

Image Courtesy of Jake Knapp, Google Ventures

Day 1: This is your empathy and assumption day. Who are your users? What do you think their needs are? What are their problems? Who might be your competitors? Ask some experts for their opinions, and then formulate some thoughts around your assumptions. What is your riskiest assumption? Which assumption might you want to test?[31]

Day 2: This is test development day. Brainstorm how you might test your riskiest assumption. Don't judge your ideas; just get them out there.[32]

Day 3: Decide on the best way to test. Storyboard (draw out) how you might test your riskiest assumption. If you're building an app, how might you lay the pages out? What prototyping medium will you use? Make a plan for your prototype and testing day.[33]

Day 4: Build your test. Build the prototype and line up customers to test it.[34]

Day 5: Test day! Use your prototype to test your ideas again and again. Test with real users from your target audience. Learn what does and doesn't work.[35]

Google Design Sprints provide you with deadlines during the discovery process. Some teams get stuck in "analysis paralysis" and don't rapidly test and iterate, but the more tests you can run with as many customers as possible, the better your product or feature will be.

What's missing from the Google Design Sprint one-week flow is enough time to do the "learn" step, which happens after the tests finish on the last day. Analyzing your test results and deciding whether you'll pivot or persevere takes time. At Capital One Labs, we found that running back-to-back Google Design Sprints wasn't sustainable because we needed time for analysis between them. At most, we would run two sprints in a row, and then take a week off to decide how to approach the next steps. Google Design Sprints are a great framework for lighting a fire under you, but they're not something you can do in perpetuity for months on end.

Discovery: Reality

At companies past the start-up stage, engineers rarely get involved in these activities. In fact, many companies don't even do a product discovery phase or incorporate these types of undertakings into their regular processes. Honestly, most companies do a lot of the things that I outlined in the beginning of Chapter 3 in the "what not to do" list.

Many companies treat their engineering department as a "feature factory." Someone on the executive team has an idea, and the product department is supposed to go out and build it. These types of organizations still function and can even make a lot of money—that's why they exist. However, this business model is a gross underuse of talent, as engineers have the unique ability to apply their knowledge of what's technically possible to solving customers' problems. You'll likely work at more than one company that functions as a "feature factory," but keep in mind that you have other options.

Discovery: Survival Tips

1. Listen to customers - If your company doesn't have robust discovery practices, you can take the first step. Talk to your customers. Ask to join your UX and product team members at customer interviews. You don't have to say anything; in fact, if the interview is online, you can sit in the room without them even knowing you're there. Or, you can ask a few questions yourself.

2. Share with your team - After listening to your customers, come back and share what you've learned with your team. Getting

everyone on the same page helps you build better products together. Better yet, record customer sessions and share videos with your colleagues!

3. Understand problems, not feature requests - Customers will come to you with feature requests, and they may even tell you exactly how they'd like them to work. Dig deeper. Try to understand the problems they're trying to solve with their feature request, and you may find a far easier way to solve that problem with another solution.

4. Ask open-ended questions - "What problems are you having with our notification reporting capabilities?" is a better question than, "Do you wish the notification reports page listed other devices?" You'll get a much better sense for what your customers want if you don't ask leading questions. If the question can be fully answered with a simple "yes" or "no," don't expect to get a lot of great information. Another thing teams often do is ask customers if they want a particular feature. That's useless! Customers will always say they want new features, but they may not be willing to actually pay for those features.

5. Put designs or prototypes in front of customers - Show customers a prototype or mockup and ask them to tell you what they're thinking as they try to interact with it. Sit silently and listen. You'll hear when they find things confusing, you'll see when they pick up what they're supposed to do right away, and you'll watch them struggle to interact with a confusing page. This technique will elicit far better feedback than simply asking questions.

6. Utilize new technology - Your superpower during the discovery process is your knowledge of technical capabilities. You can connect the dots between customer problems and viable solutions—no one else in the organization can do that better than you can.

7. Convince leaders to undertake a discovery phase - You may have to begin at square one to convince your team to try some of these discovery techniques. Start small. Ask for a week to build a prototype or proof of concept for new projects that you can share with customers. Use customer feedback to scope

and build a better system. A couple of weeks spent vetting ideas at the beginning of a project are far less expensive than finding out a feature isn't useful after completing it. "Saving time" by skipping the discovery process can end up costing the team months of work.

8. What if you can't get your organization to try a discovery phase? - You can still ask to talk to customers on a regular basis. You can organize hack-a-thons to try out new ideas; you can carve out time to build prototypes; you can sneak small experiments into your product and see how customers react to them. Take your learnings and iterate on them.

Part II: Delivery

So far, we've covered the agile principles, which form the underpinnings for all agile software development processes. We talked about what makes a great agile team and how to form that team. Then, we walked through how you figure out what to build: the discovery process. We stressed the importance of talking to customers and validating assumptions with them. Then, we outlined how to use lean start-up, design thinking, and Google Design Sprints to iterate until you reach product-market fit.

As we mentioned in *Part I*, many agile software development frameworks are out there.

The Agile Umbrella

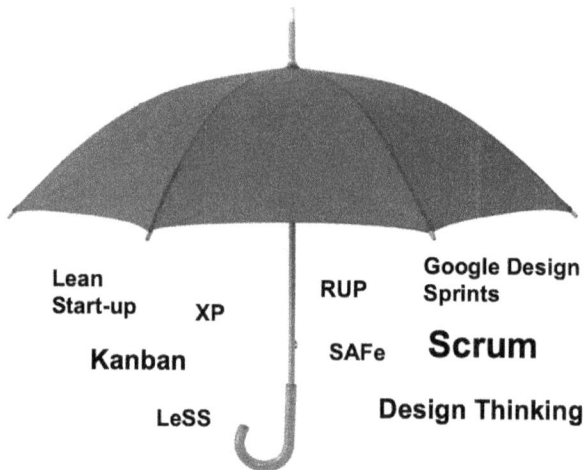

In *Part II*, we'll discuss the delivery frameworks you're most likely to encounter: Scrum and Kanban. These frameworks work best once you figure out what you're building, but many teams use them all the time, even in conjunction with the discovery frameworks.

Each framework has the same goals: to get software out to customers, **so we can provide value** (and make money). Then, we get feedback and iterate on our work. We start by translating our learnings thus far into valuable chunks of work called "user stories." Then, we organize the team to finish those user stories, test them, and get them out to customers on a regular basis.

Let's start with how we write user stories. User story writing is an important skill for both Scrum and Kanban.

Chapter 4: User Stories

U ser stories are how we write requirements for agile software development teams. They're short, simple descriptions of a feature written from the user's point of view. User stories shift the focus from writing requirements to talking about them.[1] Teams talk about user stories when they plan their upcoming work, and the user story format makes those conversations flow smoothly.[2]

User Stories: Key Concepts

User stories are created to help the team understand *what* the end-user wants and *why*. Then, the team can determine *how* to build the feature.

Why shouldn't the business dictate *how* to make a feature? Because the team knows their product inside and out. Features can be built in many ways, but only the team knows what might be the simplest and most efficient way to accomplish the end user's goals.

User stories make product requirements crystal clear. Again, they support better conversations as we build our product.

User Story Structure

User stories have two parts: the user story itself and its acceptance criteria. User stories are written as follows:[3]

As a <type of user>, I want <some goal>, so that <some reason>.

For example: *As a professor, I want computer science students to learn about user stories, so that they can write effective requirements for their projects this week!*

Or, we can imagine that a very large user story from Uber's early days might have been: *As a 30 to 35-year-old woman living in New York City, I want to be able to hail a cab at 6 a.m. and have it arrive in five minutes or fewer, so that I can get to the gym before I have to go to work.*

Note that user stories always describe a feature that's meaningful to your customers end-to-end.

Yes: *As a college student, I want to order tacos from Taco Yum for delivery, so that I can eat at home and catch a movie online with my roommate.*

No: *Add two database tables that hold taco orders and customer information.*

Acceptance criteria are bulleted statements that specify when the software is complete.[4] Each statement must be provable as true or false. Acceptance criteria make the requirement incredibly clear. When discussed as a team, they can flesh out unknowns so development can proceed.

User Story Example: *As a user, I want to create an account, so that I can shop online.*

Acceptance Criteria:

- *First name, last name, and email address fields are required.*

- *Email uses the standard format.*

- *Phone # is in US format, contains no alpha characters, and has a minimum of 10 digits.*

- *Etc.*

Note how the acceptance criteria add important implementation details to the story. Testers often use the acceptance criteria to start their test planning.

Teams may have someone in a Product Owner role who represents the business (more on this in the next chapter). If they do, the Product Owner writes the initial user stories for their team. Then, the team discusses the user stories and acceptance criteria and enhances them until they're clear enough for the team to begin its work. If teams don't have a Product Owner, they can write the stories together, or one person can volunteer to take a first stab at them. Here are a couple more examples from a handful of projects we've worked on in my class.

Budgeting Application Example: *As a high school student, I want to take a picture of my receipts and have those expenses automatically tracked in my budget app, so that I don't have to enter spending data manually.*

- *The budget app must handle long receipts (i.e., it must be able to string multiple pictures together).*

- *The budget app must be able to find and recognize the receipt total for ninety-nine out of a hundred test receipts.*

- *The receipt total must be correctly added to the budget app's expenses section.*

COVID-19 Dashboard Example: *As a college student, I want to view a map showing up-to-date information about the number of COVID-19 cases within one mile of my location, so that I can decide whether to go to class today.*

- *Students must be able to enter an address or have the dashboard glean it from their location settings.*

- *The dashboard must return the number of active cases, plus the last date/time the data was updated.*

- *Students should be able to click on the active cases number to view the daily trend in the previous two weeks.*

- *Returned data must be colored red, yellow, or green according to Health Data guidelines for high, medium, and low numbers.*

User Story Sizes

Stories come in different sizes—some are large enough to be entire projects, while others take less than a day to complete. Teams will generally start with large-project opportunities, which we call "epics".[5] Then, they break them down over time into several "features" and user stories.[6]

A user story should be small enough to be completed in one to four weeks. This is the length of a sprint, which is a concept we've defined and will discuss more in Chapter 5. The team should break down anything larger than that into multiple user stories.[7]

Let's look at an example of a product we're all familiar with, like Amazon's shopping site. Amazon's main page has widgets galore that change to capture users' attention. An epic-sized story related to the Amazon main page might read:

As a shopper, I want to see products I may be interested in on the homepage, so that I can quickly find and buy items.

- *The homepage has a "Buy Again" widget for users to re-purchase their frequently-bought products.*

- *The homepage has a "Keep Shopping For" widget showing previously searched-for items or things like them.*

- *The homepage has a "Deal of the Day" widget.*

This is quite a big story, and each of the acceptance criteria could be its own small project. These small projects are known as features. A feature is smaller than an epic but more extensive than a user story. It represents a large, valuable deliverable for a customer and takes longer than a team's one- to four-week sprint length to build. The team can break down epics into features and features into user stories. For example, the epic above may spawn the following features.

Feature #1:

As a shopper, I want a way to view my frequently-made past purchases, so that I can easily buy an item I need again.

- *The homepage has a widget entitled "Buy Again."*

- *The "Buy Again" widget shows four previously-purchased items that the user has bought more than once.*

- *Users must be able to click on each item, which will take them to the "Your Orders: Buy Again" page.*

- *The widget must have a link at the bottom labeled "See More," which takes them to the "Your Orders: Buy Again" page.*

Feature #2:

As a shopper, I want to see items I've searched for recently, so that I can continue shopping or add them to my cart.

- *The homepage has a widget entitled "Keep Shopping For."*

- *The "Keep Shopping For" widget shows the last four products the user has perused.*

- *Users must be able to click on each item, which takes them to the item's Product Page.*

Feature #3:

As a shopper, I want to see Amazon deals on the homepage, so that I don't miss out on a deal for a product I might like.

- *The homepage has a widget entitled "Deal of the Day."*

- *The widget must show the most popular, on-sale item for this user's demographic.*

- *The widget must show the picture, product name, and the percentage off, but not the product price.*

- *Users must be able to click on each item, which takes them to the item's Product Page.*

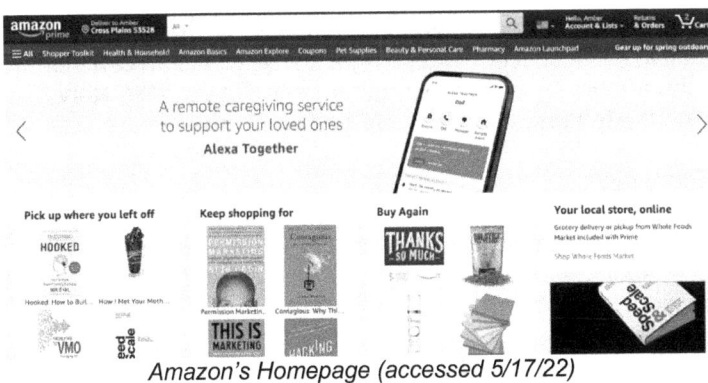

Amazon's Homepage (accessed 5/17/22)

The features are more focused, and so are their acceptance criteria.

However, many of these features are still too large and ambiguous for teams to start working on them. For example, what algorithm would you use to surface the best items in the "Buy Again" widget? Working on that logic may take a few weeks, depending on what data exists in Amazon's back-end databases.

So, features must be broken down into user stories. Each user story represents a small slice of end-to-end value, and they can be delivered to a customer within a sprint. For the "Buy Again" feature, we might see the following user stories:

User Story #1:

As a shopper, I want to view items I've bought within the last six months, so that I can easily add them to my cart and purchase them again.

- *The widget is square, white, and labeled "Buy Again" in black at the top right.*

- *The widget is located in the middle of the bottom third of the page.*

- *The widget shows four items.*

- *Each picture can be selected and takes you to the "Your Orders: Buy Again" page.*

Many user stories will include a homepage mockup. The design makes the layout crystal clear for developers.

This user story is a simple, end-to-end deliverable that provides value to the customer. When engineers complete this user story, they can release it to Amazon's homepage right away. The team can then work to refine the feature by working on other user stories. This story does not complete the feature; to do that, we need to work on more user stories.

User Story #2:

As a shopper, I want to see the relevant items I buy most frequently in the "Buy Again" widget, so that I can easily add them to my cart and purchase them again.

- *The four items in the "Buy Again" widget must be items that the user bought in the last three to six months and tends to buy every three to six months.*

- *The four items must be items the user has bought multiple times or that similar users tend to buy multiple times.*

This user story is an important algorithm enhancement aimed at serving the four products this user is most likely to buy.

User Story #3:

As a shopper, I want to see more items I've purchased beyond the four shown, so that I can quickly find the item I'd like to buy again.

- *A "See More" link is at the bottom of the widget, which takes users to the "Your Orders: Buy Again" page.*

- *Text that says, "Shop your everyday essentials" appears just below the pictures.*

- *Clicking anywhere in the widget will take you to the "Your Orders: Buy Again" page.*

You can see now how epics break down into features and user stories, which are delivered quickly to create value for customers. Then, they're iterated to enhance the previous user stories' value.

Story Organization

Stories are organized into a product backlog, which is a prioritized list of work for the team. The work ready for the team appears at the top. It's detailed, understandable, and small enough to fit into a sprint. Farther down the backlog, we see larger stories (features and epics). Over time, these features and epics break down into user stories as they get closer to the top of the backlog.[8] In this way, the product backlog looks a bit like an iceberg—the stories at the top are visible and ready to go, while the stories under the surface are larger and nebulous. They'll emerge at the surface once the team finishes the stories at the top and pulls them off the backlog.

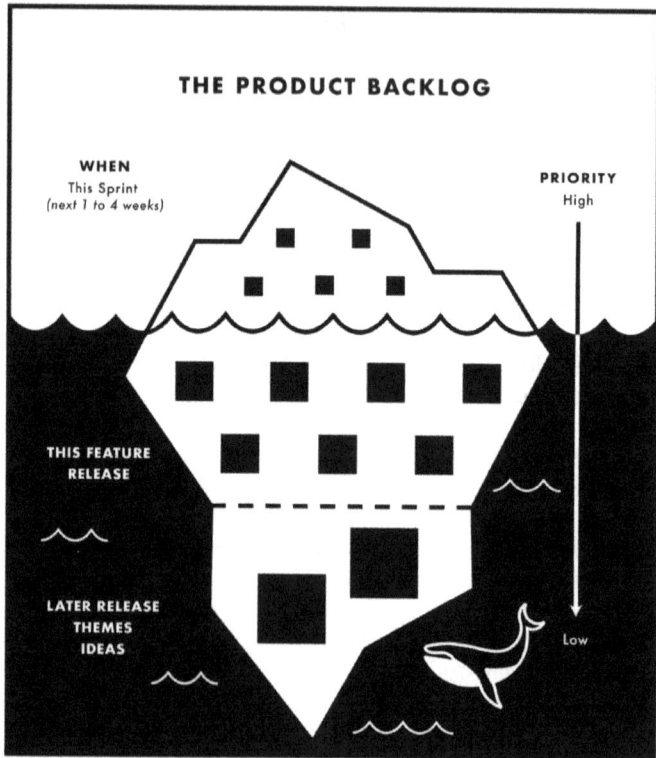

THE PRODUCT BACKLOG

WHEN
This Sprint
(next 1 to 4 weeks)

PRIORITY
High

THIS FEATURE
RELEASE

LATER RELEASE
THEMES
IDEAS

Low

User Stories: Reality

Being able to break down user stories effectively, so each one captures something valuable to deliver to a customer is a skill and a bit of an art! Most engineers develop this skill over time. It's very difficult at first, and you might hate doing it. You're also going to think it's impossible to break some stories down further, but it's almost never impossible, which is where the "art" comes in. Practice is the only thing that makes breaking down user stories easier.

Most teams also give up the official user story format. They'll create tickets in a tracking tool like JIRA, Asana, Git Tracking, or even ClickUp that likely have titles such as "create the dashboard for healthcare customers," with more details in the description field.

Teams may not work on the user stories together. Instead, they'll take whatever the Product Owner or tech lead writes and start working on it immediately. These teams miss the most important part of writing user stories: discussing them. Therefore, they miss out on potential implementation alternatives and pitfalls. These engineers will find issues when they're working on the ticket, which will slow down the work.

Finally, it's easy to forget to write user stories that represent a piece of end-to-end *value* for the customer. It's much easier to write a story for yourself for, say, the back-end code. However, that practice ignores the big picture, or the most important part of why you're writing the code in the first place: why the customer needs this work. A user story like this. . .

As an online shopper with a disability, I want to be able to click on an accessibility icon to enlarge the website text, so that I can more easily see the content.

. . .will provide a lot more value than a story that simply creates the accessibility icon itself. The functionality behind the icon is what's important. Always think about what value you're delivering to the customer.

User Stories: Survival Tips

1. "So That" - If your team isn't using the official user story format, it's still important to think about the "so that" part of your tickets. If you don't understand why you're building a ticket, ask your Product Owner, manager, team lead, or business owner. What they tell you will help you find the best solution to your customers' problem.

2. Acceptance Criteria - Acceptance criteria makes the work you'll need to do clear. If your Product Owner doesn't use acceptance criteria, work with a tester on your team to develop your own. Testers are great at thinking through the system's various edge cases; they'll benefit from understanding the work you're doing before they have to test it.

3. Breaking Down User Stories - It's damn hard, but I recommend breaking your tickets down to sizes you can complete in less than one week. Why? Because finishing work and feeling

productive is motivational. No one likes to slog along on an endless ticket for weeks or months; they like to see their work done and in production. Taking the time to break down your tickets will pay off and make the project faster in the end. Breaking down your stories also gets faster and easier the more you do it. Like most things in life, getting good at this takes practice. So, don't get frustrated if after a couple of years you still find this task difficult. You can do this breakdown together as a team. In fact, I encourage new engineers to break down tickets with more experienced developers, as the pros can teach you tips and tricks for doing so. The entire team will think of better ways to do the work. Now, you might be asking, "Doesn't it take more of everyone's time to talk about tickets together?" Not necessarily! If the team can figure out a more efficient way to do the work together, you'll save time overall.

4. Tester Demos - When you finish a ticket, I recommend you sit down with one of your testers and demo the user story to them. That way, they can ask questions, and they may even point out some things you may have missed before you turn your attention to another story.

Chapter 5: Scrum

Scrum is the most common agile framework in use today,[1] so understanding what it is and how it can be misused is advantageous. You're very likely to hear about or use Scrum throughout your career.

Scrum: Key Concepts

Scrum allows teams to work together effectively and describes a set of best practices for doing so. Its process is outlined in the *Scrum Guide* by Scrum.org.[2] We'll explore the basics here, so after finishing this chapter, you'll better understand how to work on a Scrum team. You'll also be able to start your own Scrum team if you'd like.

Ninety percent of the State of Agile survey participants in 2022 were using Scrum.[3] By 2018, it had been adopted by at least 56 percent of teams and 83 percent of software companies.[4] In a nutshell, the Scrum process looks like this:

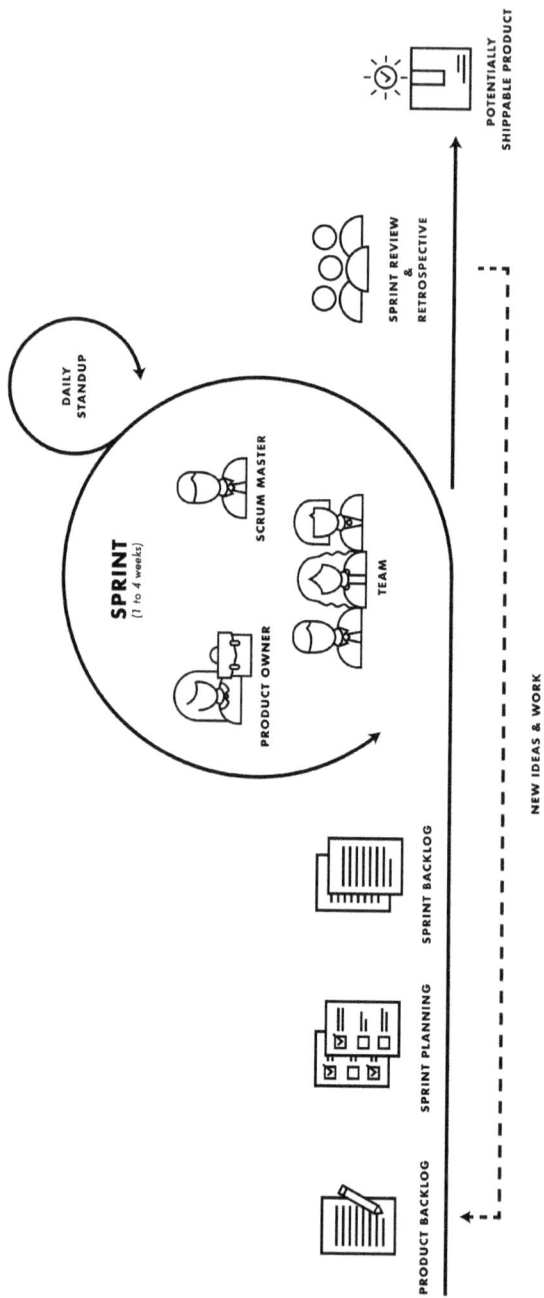

THE SCRUM PROCESS

DAILY STANDUP

SPRINT
(1 to 4 weeks)

PRODUCT OWNER

SCRUM MASTER

TEAM

SPRINT REVIEW
&
RETROSPECTIVE

POTENTIALLY
SHIPPABLE PRODUCT

PRODUCT BACKLOG

SPRINT PLANNING

SPRINT BACKLOG

NEW IDEAS & WORK

A Scrum team is small (fewer than ten people) and cross-functional.[5] Teams work in one- to four-week timeboxes called sprints, during which they pull user stories into a sprint backlog and finish the stories by the end of the sprint. Once they complete this process, they've got a "potentially shippable increment," which is a fancy way of saying they've finished a small feature that provides value to the customer. They can then choose whether to ship that feature each sprint or hold onto it and ship it every few sprints.[6]

Scrum Principles

Scrum has three pillars and five values, which are similar to the agile principles we discussed in Chapter 1.[7]

Pillars:
- Transparency
- Inspection
- Adaptation

Values:
- Commitment
- Focus
- Openness
- Respect
- Courage

Again, following the principles is the important part. Teams who care about these pillars and values tend to be healthy and successful.

Scrum Roles

In Scrum, the focus is on teamwork. The team itself consists of three official roles: Product Owner, Scrum Master, and "the team."[8] These roles should be distinct, as the Scrum Master, Product Owner, and team members are different people with unique skills and focus.

Product Owner

Product Owners are the point of contact between the team and the team's stakeholders and customers. Their job is to understand their customers' needs, the requirements coming in from all sources, and the product's market. They then translate those needs into features and user stories on which the team can work. Product Owners keep a prioritized backlog of work, so the team knows what work they can start next. They work with the team to break down user stories and answer questions about the feature along the way. When the team completes the work, Product Owners test the system to ensure the features work as expected. This process is called acceptance testing.

Product Owners have an incredibly important role. They are often called the "CEO" of their product,[9] as they are the ones who make tie-breaking decisions for the team. It can be a tough job, but it's also a rewarding one.

Scrum Master

Scrum Masters are also known as agile coaches. They protect the agile principles that underlie any good Scrum instantiation. Scrum Masters facilitate team meetings, as it's their job to remove any blockers the team hits as quickly as they can. This task helps the team move faster. Scrum Masters also shield the team from any work that doesn't fit into their main sprint goal. They'll often work with the Product Owner to ensure a steady stream of work is flowing in to the team. What Scrum Masters don't do, however, is assign work. That job belongs to the team itself because the team is self-organizing.

The Team

Everyone else on the team is known as "the team." It may consist of people with various titles and skills—you may have developers, testers, and designers on the team. It's cross-functional, self-organizing, and empowered to make its own decisions. Ideally, everyone on the team is co-located and dedicated to the team full-time (although I have yet to work on a co-located team). While it's nice to all be together in one location, in my opinion, it isn't necessary. Having a dedicated team, however, is crucial. Again, teams, including the Product Owner and Scrum Master, usually consist of fewer than ten people.

Teams get together, estimate the upcoming work, and then decide how they'll get the work done. Everyone is responsible for quality, and everyone commits to an amount of work they'll get done *as a team* over the course of a sprint.

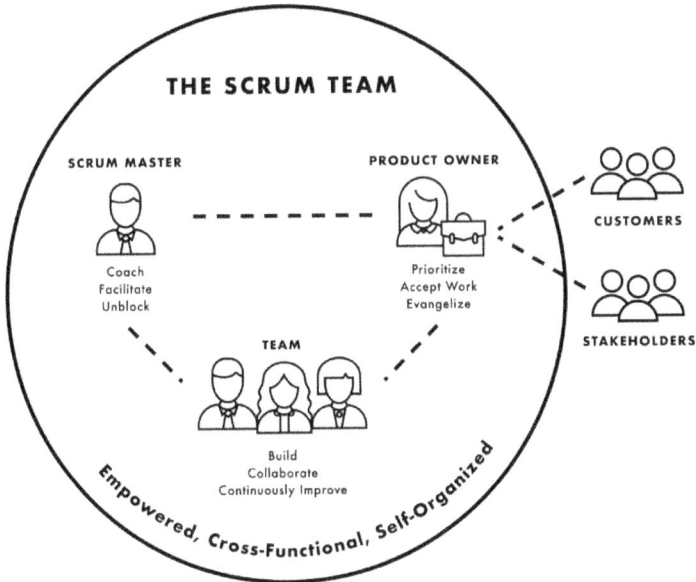

Sprints

One of the main concepts in Scrum is that work is timeboxed into something called a sprint.[10] As we've established, sprints are one to four weeks in length. Most product development teams today use two-week sprints. Service-oriented teams like IT and DevOps benefit from one-week sprints, or they can use Kanban (we'll get to Kanban, another agile framework, in the next chapter).

During each sprint, teams commit to getting a certain amount of work done, called the "sprint goal." They use the sprint length as a deadline and work together to meet their commitments. The commitment for that sprint is to complete the work to reach the sprint goal. Since other work may also be part of the sprint, if the team has to make a choice between reaching the sprint goal or finishing the other work, they choose the sprint goal *every time*. If testers need help, developers jump in; that's how important it is to finish what you've committed to do. The more versatile members of the team can be, the more the team can get done within the sprint.

Sprints are never longer than four weeks. It's easy to determine how much work can get done in one to four weeks; after that, estimating time can start to become wildly inaccurate.

No new work should be added to the sprint after it has started. This is an important concept. Teams need the chance to focus and get work done, which they can't do if priorities are constantly changing. If teams run out of work, they can pull more in, but they shouldn't add new work until they're done. In Scrum, the team's responsibility is to make sure no new sprint work shows up in the middle of a sprint. Almost nothing that arrives mid-sprint is so important that it can't wait until the next sprint starts. Most stakeholders will understand if you tell them that their work will be in next week's sprint. In essence, teams commit to the work in the sprint and they finish it. Then, they reprioritize, commit to new work, and start a new sprint. Drip by drip, this is how important projects get done—in a sustainable, predictable fashion.

Sometimes, high priority bugs will demand the team's attention right now. For example, production is down, a major feature isn't working, or a security issue arises. These scenarios should happen rarely. If issues like these come up often, it's okay to leave room in the sprint to accommodate high-priority work. Again, the vast majority of bugs can wait until the next sprint, but acknowledging and planning for the unknown can help the team meet its sprint goal.

Definition of Done

When you're closing a sprint, all your work needs to meet certain criteria, a concept referred to as the "definition of done" (DoD). The definition of done is designed by the team and is team-specific. It's a checklist that states what must be done for every user story before completion.[11] This definition doesn't only mean the work is coded and tested; it means the work can be (and in some cases is) pushed out to production and into customers' hands. Again, it's team-specific. Your team's DoD might look different than mine. Still, here's what a typical DoD might include:

- Design
- Code
- Code Review
- Testing
- Integration Testing

- Documentation
- Demo-able, Shippable Product

The team doesn't get partial credit for any work not completed in the sprint. Instead, that work moves to the next sprint, where it's completed or it goes back into the backlog (if it's no longer the highest priority work). Why don't you pat yourself on the back for a partially-finished ticket? Because you can't deliver that value to a customer. The goal is to finish each sprint with real, working, documented, production-ready code. If you don't have that, you don't get credit.

Scrum Artifacts

Scrum teams manage the work with the help of a couple of artifacts, or tools and visuals that help the team decide what to work on and whether the work is on track.

Product Backlog

Product managers are responsible for maintaining a product backlog,[12] which lists all the features they'd like to add to the product as well as bugs, tech debt, and other work the team may need to do. The product backlog is a prioritized list of user stories for the team, and they're generally kept in a ticket-tracking tool such as JIRA.

Remember our iceberg diagram from the user story chapter? That's your product backlog. The work at the top (above the surface) is smaller, refined, estimated, and ready for the team to tackle. The work farther down (underwater) is large and nebulous. It may be closer to a set of ideas than something on which a team could actually begin work. Over time, as the work moves up in the backlog, it gets broken down and better refined.

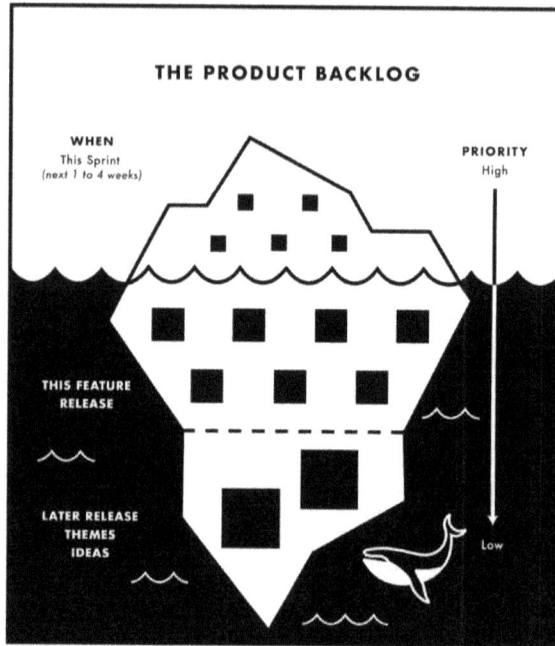

THE PRODUCT BACKLOG

WHEN
This Sprint
(next 1 to 4 weeks)

PRIORITY
High

THIS FEATURE
RELEASE

LATER RELEASE
THEMES
IDEAS

Low

Sprint Backlog

At sprint start, the team pulls the highest priority items from the product backlog into a sprint backlog.[13] They only pull in what they feel they can commit to getting done over the course of the sprint. Then, the team figures out who will work on each story, which happens during an event called "sprint planning."

Again, the sprint backlog doesn't change over the course of the sprint. No new work is added to the sprint unless the team finishes their work early, and if they do, the whole team needs to discuss and agree to the sprint backlog modification. The last thing teams want to do is miss the sprint goal to pull in an item that "just became the highest priority." You can wait to start new items until the next sprint. "Urgent new work" is almost never actually urgent. Products get built by focusing on the tasks at hand, finishing them, and then moving on to the next items.

The Sprint Backlog

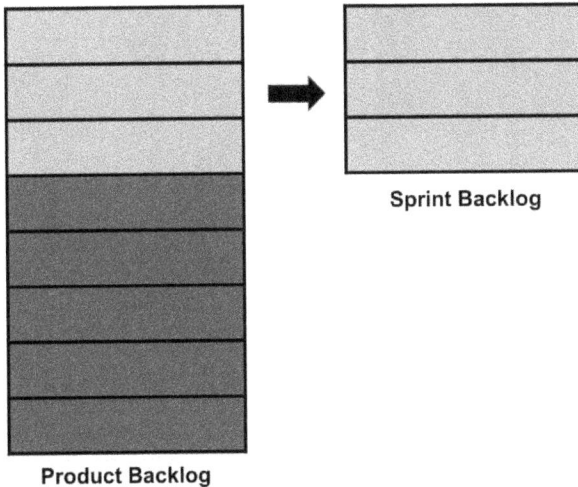

Sprint Backlog

Product Backlog

The sprint backlog is pulled from the highest priority items in the product backlog.

Visualizing Backlogs and Sprint Work

Most teams put their work into a tool designed for user story and defect-tracking. Many such tools are available, but a lot of teams use Atlassian's JIRA because it's flexible and customizable. Rally, Git Tasks, Trello, ClickUp and many others are similar, and all support Scrum.

Another way to track tickets is by putting up a board on your wall using index cards or sticky notes; however, that only works if team members are in the same location. It's amazing how satisfying a physical board can be, even though it seems antiquated. Something special happens when a team physically moves tickets around, but I doubt you'll encounter this technique on most of your future teams—we're too distributed these days.

Which tool you use isn't important. What's important is that the team has somewhere to go to visualize the backlogs and see where the work in progress is. Most teams need three visuals throughout the sprint.

1. Product Backlog
2. Sprint Backlog
3. Sprint Work in Progress Board

The product and sprint backlogs might look something like this, especially if you've been backlog refining and know what might land in the next couple of sprints:

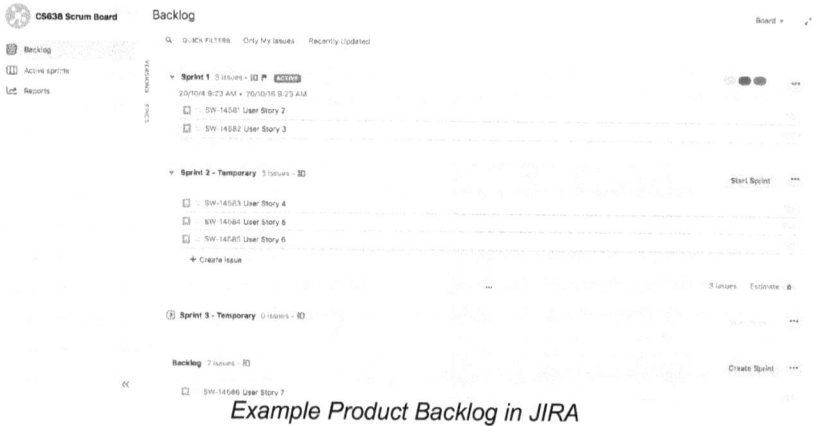

Example Product Backlog in JIRA

Here's an example of a sprint backlog from JIRA:

Sprint Backlog in JIRA

Once a team is working on a sprint, they need to see what phase the work is in. Phases might be "development," "blocked," "testing," or "done." Teams also look at who owns the work and how much work is in progress across the team. To do so, teams usually use a Kanban-style board.

Kanban is the Japanese word for "sign" and is the next agile framework we'll discuss. Teams use a Kanban board to lay out exactly where each user story is in its lifecycle and how much work is in progress. It looks something like this:

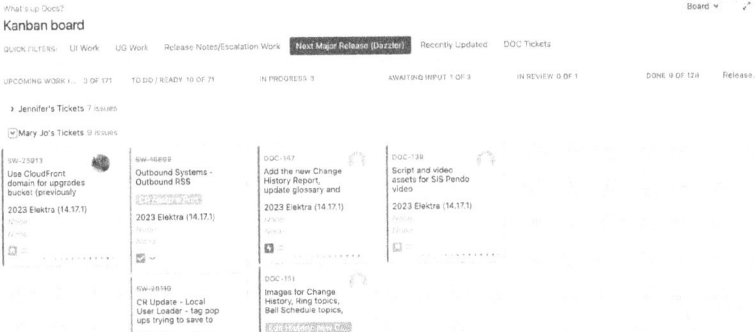

A Kanban-Style Team Board in JIRA

The columns represent each development phase. A team's board might have columns called "to-do," "design," "development," "testing," "documentation," or "blocked," plus any other phase important to the team. Work is pulled into the next column when a team member is ready for it. Everything moves across the board from left to right until it's done, and owners change when appropriate. In this way, the Scrum Master (or anyone else on the team) can see, at a glance, what's happening with each ticket. This is the view teams review and discuss every day during an event called a "daily stand-up."

Let's talk a bit more about the events in each sprint that support reaching that sprint goal commitment.

Scrum Events

Scrum recommends all teams have a few touchpoints. Scrum calls these "events," but most people think of them as meetings.

Stand-Up or Daily Scrum

Each day, the team holds a daily stand-up, or daily scrum, to check in on where everyone is with the committed work.[14] The daily scrum is fifteen minutes long or shorter. You can successfully run this meeting in a couple of ways.

The *Scrum Guide* now suggests that teams focus on the work during a stand-up. They look at their sprint backlog or sprint board and talk about each user story. They point out what's done, what's left, and whether any blockers are inhibiting work completion. This agenda helps the team focus on the work at hand rather than the team members themselves.

In the *Scrum Guide*'s past iterations, the meeting agenda was to ask each person three questions:

1. What did you do yesterday?
2. What will you do today?
3. Do you have any blockers?

If anyone's work is blocked, the team tries to quickly unblock it, so the sprint work can continue. The Scrum Master's job is to help unblock the team members. Many teams still use this meeting agenda for their daily stand-ups.

Regardless of how you approach it, the daily stand-up should take no longer than fifteen minutes. The most common reason for a stand-up to take longer is that teams try to solve issues or discuss work at length. The Scrum Master should recognize when this happens and stop the discussion because teams can save in-depth topics until the meeting ends. Only the people involved in the topic need to stay; the rest can leave and get on with their work.

Menlo, Inc., a company in Ann Arbor, MI, has adopted many forward-thinking practices. They allow people to bring their kids to work (even babies!) and always work in pairs (also known as pair programming, which we'll discuss in Chapter 7). They also have a daily stand-up for their whole team of fifty people. Yes, fifty. Every morning, everyone shares, and it still gets done in fifteen minutes or less. It's not a miracle; they're just disciplined—they save all their discussions for separate meetings with the right people.[15]

Sprint Planning

Sprint planning occurs at the beginning of the sprint. During sprint planning, the team pulls a small number of the highest priority items from the product backlog into the sprint. They stop when they feel they've extracted enough work for the team in the sprint. The team also discusses implementation specifics like what actually needs to be done, who might work on it, and how long it might take. In that way, work may be broken down into smaller pieces. Design discussions take place, so the team can begin working on it immediately. The team covers all these points in real time during the sprint planning meeting.

Once the team has enough work for the sprint, they commit to a sprint goal and to getting that work done. Then, they start. The desired outcome is a shared understanding of the work and team commitment to getting it done.[16] Again, once the sprint starts, nothing new gets pulled into the sprint backlog.

Backlog Refinement

Backlog refinement doesn't show up in the *Scrum Guide* as its own event, but many teams find it essential to running the team well.[17] Once a week, or once every sprint, the Product Owner runs a session to look at the upcoming work for the next one to three sprints. He or she breaks down the work to ready it for the team's sprint planning. The whole team can take part in backlog refinement, or you can refine the work with only the Product Owner, lead developer, and lead test engineer. When in doubt, include the whole team. Using the smaller group allows the rest of the team to focus on its sprint work, but they still need to understand the work, so they'll need to be educated during sprint planning.

The backlog refinement agenda looks like this:

1. Look at the first user story in the product backlog. The Product Owner explains the work to the team and answers questions.
2. The team discusses exactly what needs to get done and may break the user story into smaller pieces.
3. The team estimates the work and then documents the discussion in the ticket or tickets.
4. Repeat until you run out of time, or you've discussed enough work for the next one to three sprints.

Again, this process can happen as a separate meeting or as part of sprint planning. When you run backlog refinement is up to the team. If you do this as a separate meeting, once a week for an hour or once every two weeks for ninety minutes to an hour works well.

Sprint Demos

At the end of the sprint, the whole team gets together and everyone shows off the work they've done. No preparation is necessary for this meeting—no PowerPoint slides, videos, or any other theatrics. Teams get together, pull up the features they've been working on, and show how they work. If you only worked on back-end code, that may mean bringing up a console screen or even the code itself. Showing front-end (user interface) work is more straightforward.

The entire team, including the Product Owner and UX lead, should be at the meeting, as this is the team's chance to discuss any changes that need to be made or anything that doesn't look quite right. Any changes the team agrees to should be added to a ticket, and that ticket goes back in the product backlog for a future sprint.[18]

Demos are a great way to help teams meet their commitments—when you know you have to show your work at the end of a sprint, you try harder to get it done! It's also a great way to get valuable feedback from team members before releasing work into production.

Sprint Retrospectives

Once the demo is complete, the team runs what may be the most important event of all—the retrospective. Retrospectives are exactly what they sound like: a time for the team to reflect and talk about what went well, what didn't go well, and what might need to change moving forward. The retrospective allows them to discuss wins and sore points, and then fix them before they become real problems.[19] This is the practice that best drives continuous improvement.

Technically, teams should run retrospectives each sprint, although I've seen successful teams elect to run them less often, like once a month. Teams may also run retrospectives after each release.

It's important for *only* the core team to be present for the retrospective. These meetings need to be a safe space in which the team can be fully honest with one another. When you introduce managers, you may hesitate to broach subjects that need to be covered because almost no one likes to throw teammates under the bus. Retrospectives are a place where people feel psychologically safe and share whatever is on their minds.

The agenda for the meeting is:

1. Discuss what went well.
2. Discuss what didn't go well.
3. Talk about what you can do about the things that didn't go well.
4. Come up with at least one action item to carry forward into the next sprint.

During the course of the retrospective, you may find dozens of things you'd like to change. However, that much change is often too much for a team to handle all at once. Focus on one or two things that will really move the needle. Get those working well, and then fix the others after a future retrospective.

I generally use a Google Sheet to run my retrospectives. The first five minutes is for the team to silently add items into the spreadsheet, so we don't have a lot of groupthink. Then, I take the team through the spreadsheet out loud. Items entered become a blueprint for the conversation.

What went well?	What didn't go well?	What should we change?	Owner
I like how we're following the scrum process	We learned Alex is leaving -2 :(	Clone Alex	
Virtually done with React!	Designs are a bit in limbo because we need more information about the resource	This is to be expected given that we all need to learn and this is our biggest resource. Build a checklist	Mitch

A Google Document Retrospective Example

You can run a retrospective in a lot of fun ways. For example, in the Sailboat Retrospective,[20] the team draws a big picture of a sailboat on a whiteboard (you may find that your teammates have hidden talents!). Then, you talk about the wind filling the sails (what went well) and the anchors weighing it down (what didn't go well). You can have the team point out the rocks in our way (risks), and remind each other of the beautiful places they're going (the goal!). When you run retrospectives after every sprint, it's nice to mix things up a bit, and you can find many more ideas to do so online.

Sailboat Retrospective Example

Sprint Calendar

Putting all these events together, you get a good sense of what the team is doing on any given day during a sprint. A typical two-week sprint calendar might look something like this:

	Monday	Tuesday	Wednesday	Thursday	Friday
Week 1	SPRINT START Sprint Planning (2-4 hrs)	Daily Scrum (15 mins)	Daily Scrum	Daily Scrum	Daily Scrum
Week 2	Daily Scrum	Daily Scrum Backlog Grooming (1-2 hrs)	Daily Scrum	Daily Scrum	Daily Scrum Sprint Review (1-1.5 hrs) Sprint Retrospective (30-60 mins) SPRINT FINISH

Sample Sprint Calendar

Notice that on most days, teams spend only fifteen minutes in a meeting. They spend the rest of the time working on the commitments they made in sprint planning.

Scrum Metrics

Three metrics are useful in tracking agile teams: work in progress (WIP), velocity, and burndown. That age-old saying, "You can't improve what you don't measure" is not exactly true. The retrospective offers you a wonderful way to continuously improve too, and these metrics can help point out where things aren't quite right, so teams can discuss them at their retros. Let's take a look at each one.

Work in Progress (WIP)

Work in progress (WIP) is how much work a team has started but not finished. The more WIP a team has, the longer it takes to get that work done. Teams all have many competing priorities, so efficient teams limit how much they work on at once and focus on one or two items per person at most.

Measuring WIP is essential for raising throughput,[21] which is how much the team gets done. When people can focus on user stories in a serial fashion, they get their work done faster.

You can use the Kanban board we talked about to help measure WIP. Count the number of tickets you have in each column between "to do" and "done" (this is your work in progress). If the number is greater than the number of team members x 1.5, that team is likely moving slower than they could. We'll mathematically prove this concept is true in the "Kanban Principles" section of Chapter 6.

To combat having too much work in progress, start by adding WIP limits to your board. Each column can have a limit on how many tickets can be in the column at any one time. Lower that number to 1.5x the number of people on the team, and see how that works. Adjust up or down as necessary.

Why shouldn't the limits be 1x the number of people on your team? Isn't that the amount of focus we all want to have on our projects? Yes, but sometimes our work gets blocked, so it's nice to have something else to work on while we wait to be unblocked.

Measuring and limiting your WIP is one of the best ways to help a team become more efficient.

Velocity

Velocity is a measure of throughput, or how much work a team can get done in each sprint.[22] It is generally measured in story points or the number of tickets the team completes during that sprint.[23] We'll cover story points in the next section; they're a relative measure of estimation.

When teams start tracking velocity, they'll find it can be wildly different each sprint. In one sprint, they may get thirty-four tickets done, while in the next they only get to fifteen. After around four to six sprints, however, velocity tends to hit a plateau and stabilize. Once that happens, you can use a team's velocity to estimate how much work the team will get done each sprint. Teams can extrapolate that estimation to figure out when they might deliver a project.

Velocity is specific to a given team. Each team may stabilize at a very different velocity. Comparing velocities across teams is never fair, as individual teams estimate differently and break their tickets down to different sizes. Velocity should only be used as a measure of throughput for that specific team and as a measure of maturity. If a team's velocity doesn't stabilize, the time has come to examine why and make some changes.

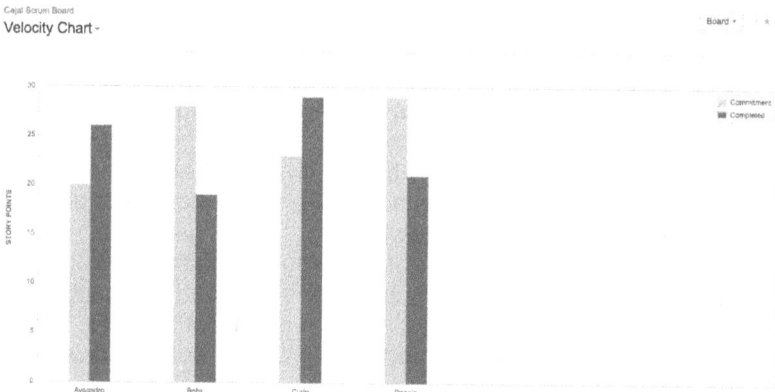

Example Velocity Chart from JIRA

Burndown

Burndown is a measurement of how sustainably the work in a sprint is getting done.[24] It tracks how many story points or tickets are closed each day during that sprint. A perfect burndown chart looks like a set of stairs or a line headed down and to the right, as it shows that the team is completing work each day.[25]

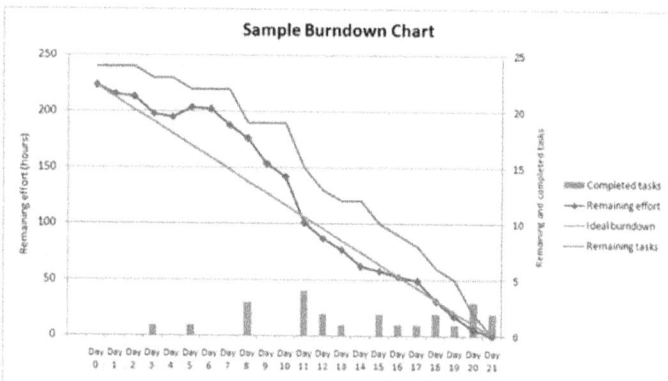

Image Courtesy of Pablo Straub

A bad burndown chart looks like a straight line with a steep drop-off at the end of the sprint. Or worse yet, it's a line that goes up, indicating that you added more work to the sprint after it started.

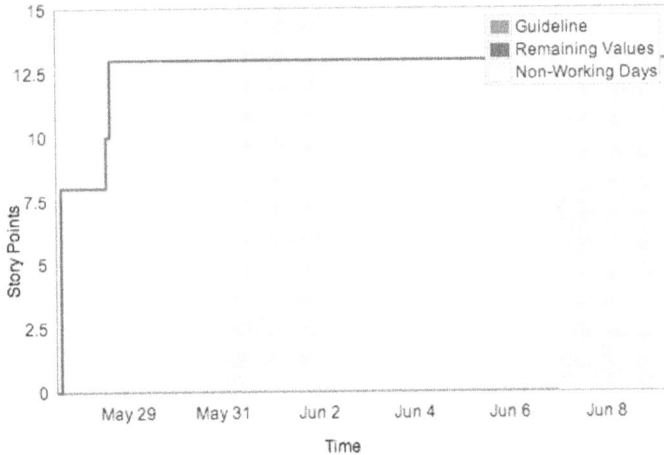

Bad Burndown Chart Courtesy of agilecoachjacque.com

Why are burndown charts important? Well, as we learned with WIP, team members should be working on one item at a time. Ideally, teams start with the highest priority work in the sprint, finish it, and then start a new ticket. The more WIP a team has, the longer the work takes, and the higher the risk is that they won't finish their sprint commitment. Burndown charts give the team and Scrum Master a sense of how the sprint is going. Watching burndown, the team can make adjustments early before they miss their sprint commitment.

Estimation

Estimation is a necessary evil. I take estimates with a grain of salt because they're always wrong. It doesn't matter whether you're a seasoned developer with twenty-five years of experience or a new hire right out of college—estimates are wrong. They're simply guesses, and we ought to treat them as such. The best way to go about making an estimate is to use the fastest, most accurate method you can. You know it'll be wrong, so do the best you can in the shortest time possible. The entire team should be involved in making the estimates, so everyone owns them and feels accountable to them.[26] The two types of estimates are absolute and relative.

Absolute Estimates

Absolute estimates are fairly straightforward. How much time will this work take?

- Four hours?
- Two days?
- Five months?
- A year and a half?

Absolute estimates are time-based. Teams estimate the time it will take to complete each piece of work, and then they add them all up and come up with a project timeline (which, again, will be wrong). In many organizations, upper management will internalize that estimate, assume it's infallible, and plan everything around that date. Then, they'll get upset when the team inevitably misses their estimate. It's better to acknowledge that estimates aren't perfect and work with time ranges. For example, say, "This work will take two to four months" and/or work with relative estimates.

Relative Estimates

Relative estimates don't have a time-based component at all. Instead, teams look at all the work before them and estimate it relative to all the other work they have to do. To start, they find the smallest, easiest piece of work they have, and then base all the other estimates on how much larger that work is than their smallest piece.

The team gets to decide what values to use for estimating and agree upon those values. Two of the most common sequences teams use are T-shirt sizes or the Fibonacci sequence.

T-Shirt Sizes

Example T-shirt sizes look like this: *S, M, L, XL, XXL, etc.*

Your smallest unit of work would be labeled as "S" for small or "XS" for extra small. Work that's a little larger would be a medium (M), work larger than that a large (L), and so on. T-shirt sizes work well because most people have a frame of reference for how much larger a medium is than a small. We've worn clothing all our lives and the sizing seems intuitive.

Fibonacci Sequence

You might remember the Fibonacci sequence from a math class long ago:

0, 1, 1, 2, 3, 5, 8, 13, 21, 34…

Each number is the sum of the two numbers that came before it. When estimating product work, engineers tend to simplify this range of numbers and limit how far up they can go. A common range might be:

1, 2, 3, 5, 8, 13

Anything over thirteen is so big that it must be broken down more before it can be estimated. Using the Fibonacci sequence method, your smallest piece of work is labeled with a "one." Anything approximately twice as large is a "two," three times as large is a "three," and so on. The Fibonacci sequence gives us more separation between the various sizes of work. The larger the work, the more error is likely to be in its estimate, so the larger its value relative to the others.

Story Points

We have a name for the relative, numerical estimates we use on agile teams; they're called story points. Story points have no time-based equivalent. Many teams try to translate story points to time ("one" is equal to four hours, "two" is equal to a day, etc.). However, those teams are missing the point (pun intended). Remember: Estimates are always wrong. Take the time-based approach away, and we're forced to talk about timelines in the abstract and as ranges, which is a much healthier practice.

Teams can measure their velocity based on the number of story points they complete each sprint. Over time, team velocity stabilizes, but it usually varies by a few points here and there during each sprint. Once it does, you're able to understand the range of time over which a project might be completed. Take the total points left, divided by the average number of points completed each sprint, multiply that number by the number of weeks in a sprint, and you've got a guess for how long a project might take. Present that number as a range (plus or minus a sprint or two on either side). Now, you've got a decent understanding of when a project will be done.

Teams usually make their estimations during backlog refinement and sprint planning. Many companies do a rough round of estimation at the beginning of a project, which gives them a sense of how long everything will take and helps them decide whether to move forward. Teams can use relative estimation wherever they estimate, and the more they break down and examine the work, the more accurate the estimates will be. Still, the point bears repeating: Estimates are always wrong.

No Estimates

That truth leads me to touch on not doing estimates at all.[27] Once teams get good at breaking down their work, most stories will be approximately the same size, and if that's happening you don't even have to estimate the work. Count your tickets and use the sum as a measure for velocity. This strategy is simple, quick, and gives your team a decent idea of how long work will take.

Planning Poker

Planning poker is a fun technique that leads to some pretty great estimating discussions. The method was created by James Grenning and popularized by Mike Cohn in his book *Agile Estimating and Planning*,[28] and it works as follows:

Once you've agreed upon a set of values for estimating, create a deck of cards for each team member, with one possible value on each card. You can buy planning poker cards, especially if you're using the Fibonacci sequence. The deck may have the values 1, 2, 3, 5, 8, 13, 21 all the way up to one hundred. You'd give a deck like that to one team member.

Whenever you estimate as a team, pull out your decks of planning poker cards. The Product Owner explains a given user story until everyone understands the work, and then team members secretly select a planning poker card that represents their estimate. All team members reveal their cards at the same time, so everyone can then see everyone else's estimates.

At this point, the likelihood of everyone agreeing on an estimate is slim. You'll probably see quite a variety of story points on the table, and that's okay. The point of the exercise is not to get to a value in record time; the point is to drive a conversation around how much work the team feels this user story entails.[29]

The Scrum Master or Product Owner might ask the person who put down the lowest value why they chose that value. They would likely do the same thing for any other outliers as well, and during this conversation, the team may find out some new information. For example, a user story might require the test team to build a whole new testing infrastructure, which would increase the estimate. Or, maybe someone has already written code that could be reused for this user story, which could also reduce the estimate. Once everyone has had a chance to explain their scores, the team does another round of planning poker. They pick their new estimate and then reveal their cards at the same time again. With each round, the team should get closer and closer until they can agree on a single value for that ticket.

In some cases, the team never arrives at a complete consensus. Most members may still think the estimate's story point value should be a five, for example, while one person thinks it's a thirteen. If that person has been heard, and the majority of the team still feels it's a five, the team would land on that value.

Some team members immediately see the value in planning poker, while others are skeptical. I admit it sounds a bit gimmicky, but most people like it once they try it. They appreciate the conversation it drives, and that's the whole point: to include everyone on the team and discuss the work at hand.

Scrum: Reality

In reality, a lot of half-assed instantiations of Scrum are out there, but a lot of good versions that work as well as the "ideal" version described above also exist. You'll meet engineers who hate Scrum or agile and anything connected with the two. If you find yourself on a team that hates Scrum, then look at Kanban (Chapter 6). It's much easier to master and see value right away.

Roles

I've worked at several companies that form a Scrum team without the proper roles involved. Or rather, the people don't understand their new roles. For example, they take project managers and anoint them as Scrum Masters, but these people keep doing basically the same things they were doing before. Scrum only works when your Scrum Master understands the agile principles. They exist to unblock the team and make sure they're communicating. Product Owners also come in all kinds of flavors; some do a great job of working with stakeholders and the team, while others don't. Many companies have shared Product Owners and Scrum Masters who manage multiple teams. When that happens, you're far more likely to find that neither of these people have enough time for your team. Some companies don't even have the Product Owner role.

Sprints

Many teams use two-week sprints. On teams new to Scrum, the committed work doesn't always get finished within that sprint, so they move the unfinished work to the next sprint. Or, they'll split the work into two tickets, so they still get credit for some work in the prior sprint. Work may not get done for many reasons. For example, perhaps it isn't ready to go at the beginning of the sprint, or maybe new work is entering the sprint in the middle. Perhaps the team finds out something is harder than expected. The Scrum Master's job is to get to the bottom of why the team isn't finishing its work and to correct it.

Teams may not deploy to production at the end of every sprint. Instead, they hold onto code for a few sprints before releasing it to production. This method isn't bad, per se, but teams commonly hold onto work until larger features are finished. A healthier practice is to deploy to production behind a feature flag, so customers don't see the new feature until it's finished. You'll learn more about feature flagging in Chapter 7.

Artifacts

The product backlog can be a mess unless you've got a very organized Product Owner. I've seen engineers or the Scrum Master step in to organize the backlog, which can work well if they have a good working relationship with the Product Owner.

Events

Some teams don't meet every day for stand-up. Instead, they'll meet every other day, or give each other updates on slack. Teams may also end up waiting for a couple of sprints and doing retrospectives less often, and they may modify other events as well. However, this is one area where cutting corners tends to be a slippery slope. Event compromises can lead to failing Scrum instantiations. We'll cover ways to skillfully change Scrum events in the "Survival Tips" section.

Giving extra attention to how teams treat retrospectives is worthwhile, as it's common to see teams ignoring them altogether. Why do they do this? Because it's easy to get to the end of a sprint and decide things are going pretty well in general. If that's the case, then why should you spend time as a team talking about improvements? So, the team skips one retrospective, and then another. Before you know it, processes get stale. Teams that skip the retro process miss out on a lot of amazing, continuous improvement.

Metrics

The main metric teams track in Scrum is velocity, but sometimes they get lazy about tracking anything. Various tools provide reports that help teams track everything from velocity to throughput. The reports are there; teams need someone to go out and look at them!

Estimation

I'll say it yet again: estimates are always wrong. The point of estimating is to have a conversation about the work, so the team is better prepared to do it. From a business standpoint, the point of estimating is to know when projects will finish. Almost every organization has their own process for estimating, but they're all inaccurate. We'll talk about a couple of tips for how to get through estimating in the next section.

Scrum: Survival Tips

1. Events (Meetings) - If Scrum seems a little meeting-heavy, you can combine some of them. My teams treat the last day of one

sprint as the first day of the next sprint, so they hold a team review (demo) followed by a retrospective. Then, they take a short break and roll right into sprint planning. You can also combine backlog refinement and sprint planning, as long as your team doesn't take an excessive amount of time to plan. If any of these meetings end up going longer than one or two hours, you may want to decouple them again for your own sanity.

2. Backlog Refinement - Speaking of backlog refinement, it's a good idea to do this at least once per sprint. If you hate slogging through the backlog, the whole team doesn't have to participate. The Product Owner, lead developer, and lead tester can take part, which will allow other engineers to keep working on other items. Everyone will find out what's coming up during sprint planning.

3. Stand-Ups - Holding these every other day or over slack can work. The stand-up schedule is worth experimenting with once your team is high-performing.

4. Blockers - Don't wait until stand-up to bring up blockers. Charlie Ciccia, Director of Development at Singlewire offers a good rule of thumb for how long to struggle with something—he says struggle for twenty minutes. If you can't come up with any ways to solve the problem at that point, ask for help. If you have ideas, spend another four hours trying to solve the problem. If you haven't solved it by that point, ask for help.

5. Retrospectives - It's tempting to skip retros, but DON'T DO IT! Even when things are going well for a team, you can always find something to improve. Building the habit of holding retrospectives creates high-performing teams. Teams that skip their retrospectives are definitely at a disadvantage.

6. Automation - If you're doing the same thing more than twice, take the time to automate it. Your future self will thank you.

7. Help Your Scrum Master Help You - Meetings can take over when they aren't timeboxed, which is why stand-ups are set at a strict fifteen minutes, regardless of the team size. If someone goes on a tangent, point it out and suggest they talk about it after the stand-up. Your Scrum Master will be happy to have

the reinforcement.

8. Product Owners - This role is key, and unfortunately, good Product Owners can be hard to find. If yours is missing, too busy, or if you have more than one, that's a problem. If yours doesn't listen to the team or stakeholders or doesn't attend meetings and keep your backlog clean, that's also a problem. Work with your Scrum Master, team lead, and manager to find a solution.

9. Non-Scrum Meetings - You'll get pulled into other meetings that aren't outlined in the events section. When you're invited to a meeting, ask yourself what you need to get out of it. If the answer is that you need to know what was decided or to read the notes, decline it and check back in later. If the answer is that you truly need to be an active part of the discussion, and it's a high priority for this sprint, then attend the meeting. Try to clump your meetings so you have plenty of open space in your schedule for deep thinking and coding work. I recommend blocking time off on your calendar and defending it with your life. That way, you know you'll have the space to finish important work during the day.

10. Meeting-Free Days - I'm a fan of meeting-free days. Your managers and leadership team may not be because they want to talk to you, but they also want you to get your work done, and they can't have it both ways. So, if you can get your whole team to agree to an entire day per week where you can focus on code, do it!

11. Nights and Weekends - Create boundaries with your time. Find companies to work for that don't have a 24/7 culture. If you hear people bragging about how many hours they worked last night, run. Burnout is real, and what many Americans don't realize is that your productivity actually goes up when you have time to rest. In my early career, I didn't get the performance rating I wanted one year at IBM, even though I worked insane hours on our projects. So I got angry and decided to stop working more than forty hours a week. What happened to me next was amazing: The next year, I did get the top performance rating. Not only that, but I was well-rested and had learned the important art of prioritization. I've rarely worked overtime since, yet I've enjoyed a fulfilling career filled with promotions and

raises.

12. Deadlines - Deadlines are pretty useful, which is why sprints exist—they give us artificial deadlines, so we force ourselves to keep making progress. Get in the habit of meeting sprint deadlines and you'll become very productive.

13. Scrumban - Many teams combine Scrum and Kanban (the subject of our next chapter). This combo-framework can be pretty effective and lighter-weight than Scrum. Check out the Scrumban section in Chapter 6 to learn more.

14. Estimation - Have I said it enough yet? Estimates are always wrong! Do the easiest thing that will give you the most accurate estimate. I like counting tickets, myself. The fact that not all projects need to come in on time is also worth noting. Marty Cagan wrote two books on product management, called *Inspired*[30] and *Empowered.*[31] He calls important projects "high integrity commitments." These are the projects sales and marketing are dying to sell, so focus your efforts on making high integrity commitment estimates as accurate as possible. The rest of your company needs you to deliver these projects on time, whereas the other projects aren't as important and can shift to accommodate high priority projects.

Chapter 6: Kanban

I think of Kanban as the gateway drug to agility. It's a simple way for teams to adopt agile practices and principles, and it doesn't have as much set-up and overhead as Scrum. It also works great for teams with smaller, varied projects coming from different stakeholders. Teams like IT, DevOps, UX, or even marketing tend to love Kanban.

When it comes to processes, I always like to have "just enough." Kanban, for many teams, is just enough process to make their team far more effective.

Kanban: Key Concepts

In this section, we'll look at how Kanban works and then compare it to Scrum.

Kanban is a Japanese manufacturing system that regulates production flow using an instruction card that's sent along the production line.[1] In software, we can also take advantage of this concept. User story work flows from one step to the next along a continuum (from design, to development, to testing). Teams need to know where that work is on the continuum at any given time, so Kanban introduces a visual (Kanban board) and the practice of limiting work in progress (WIP) to help teams collaborate and get work done faster.[2]

Kanban Principles

Little's Law is a handy queuing formula which proves that work delivery takes longer when we have too much work in progress (WIP).[3]

Cycle Time = WIP / Throughput

Cycle time is the time between a task's initiation and completion. The lower the cycle time, the faster your features get into customers' hands. *WIP* refers to the entire team's average work in progress. You generally

95

count tickets in progress to determine our WIP. *Throughput* is the number of tickets getting done over a period of time. Let's take a look at an example:

You're an engineer trying to improve how quickly your team gets work done. The team has a *WIP* limit of nine, and its *throughput* each day is three.

$$Cycle\ Time = 9/3 = 3\ days$$

Using Little's Law, we discover that it will take an average of three days to get any individual task done for this team.

So, what's the easiest way to improve how quickly work gets done? Make your team work overtime? Add people?

Let's not go crazy here. Instead, let's play around with the formula a bit. Let's say you lower the *WIP* limit to six and the *throughput* remains at three.

$$Cycle\ Time = 6/3 = 2\ days$$

By making a very simple change, the whole team is able to shave a day off of every task they touch! What if you went in the opposite direction on *WIP* (WIP = 12)?

$$Cycle\ Time = 12/3 = 4\ days$$

Well, that's obviously worse! **How quickly you deliver work is directly impacted by how much work you have in the system at any one time!**

Remember the analogy we discussed around rush-hour traffic? Fewer cars on the highway can obviously go faster than when traffic is congested. WIP limits do for teams what variable-rate HOV lanes (carpool lanes where the toll goes up when traffic increases) do for traffic. As toll rates go up, fewer people take the carpool lane, so as the traffic decreases, the cars can go faster. If the amount of work in progress decreases, the team can increase their speed as well.

Simple, elegant, easy. It works in life too. If you're feeling overwhelmed, reduce the number of things you're doing and focus on getting the items that really matter done.

Kanban Artifacts

The team's Kanban board has columns representing each stage of work a team goes through to complete a user story. We introduced the Kanban board in the "Scrum Artifacts" section in Chapter 5. The board can be complicated, showing each stage in the "definition of done." Or, it can be as simple as three columns: "to-do," "in progress," and "done." Most teams end up with a board like this one:

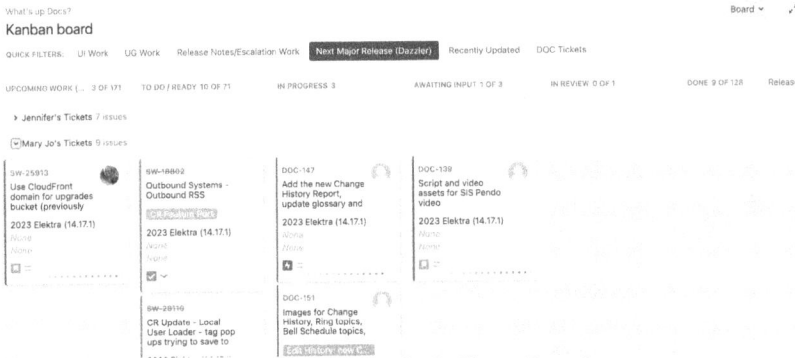

Sample Kanban-Style Team Board

Common columns also include "blocked" and "documentation." Any stage your team feels is important should show up on the board.

Cards that represent a user story populate the board and move from left to right. Team members pull the work into the next column when they're ready for it. Work in progress (WIP) limits dictate how much work can be in one column at a time. WIP controls the flow and optimizes the team's throughput.

When using Kanban, it's important to apply WIP limits to every column except "to-do" and "done." When you're starting out, a good WIP limit for the system is usually 1.5 to two times the size of the team. If you have three developers, the WIP limit on the "development" column is somewhere between four and six, which allows individuals to pivot if they get blocked but still focus and get things done.

Kanban Events

In its simplest form, Kanban doesn't include any meetings. Team members move their work across the board at their own pace, and they touch base in an ad hoc way. In reality, though, most teams need a bit more formality to perform at their highest potential. A team that already works well together can start using a Kanban board tomorrow, and they'll immediately benefit from the added visibility, with or without meetings.

If some work progresses slowly or falls through the cracks, hold a stand-up. You can do this twice a week to start, and the team can look at each item on the board and discuss and/or move it if necessary. A meeting every week or two to look at the items in the "to-do" column may help too. Use it to rearrange priorities and discuss how tickets might get completed, similar to a sprint planning meeting in Scrum.

Kanban Metrics

Four main metrics are valuable for teams to track using Kanban.[4]

Lead time is the time from when an item hits the "to-do" column on your Kanban board to when it's delivered to production. It's a valuable measure for how long something takes to be delivered. Lead time can help you estimate when a new project might be finished, and it takes into account all the other work in your queue, not only a specific ticket.

Cycle time is the time from when a team starts to work on a user story to when they deliver it. The moment that item gets pulled out of the "to-do" column on the Kanban board, the cycle time counter starts. If you know your cycle time and WIP, you can use Little's Law to calculate your throughput.

Throughput is the measure of how much work is completed during some set amount of time. Many teams using Kanban might calculate throughput for each week, for example. You track how many tickets or how many story points a team finishes each week, like Scrum team velocity. Over time, a team's throughput will stabilize and can be used to predict when work or entire projects will be completed.

We've talked at length about *work in progress (WIP)* and WIP limits in this section already. Keeping tabs on your work in progress will improve a team's cycle time and throughput. If you only track one thing, make it WIP.

Scrum vs. Kanban

Most teams will already be using Scrum or Kanban when you join them. If you have the luxury of introducing one of these frameworks to your team, think about the following topics, as these considerations will help you make the best choice.

Roles

Kanban doesn't have any special roles. Scrum has the Scrum Master and Product Owner, although some teams work great without them. Product Owners do the work of talking to stakeholders and organizing the backlog, whereas Scrum Masters help the team adhere to the agile principles and become unblocked. If you like the idea of having people concentrate on those areas so you can code, consider Scrum. If you function pretty well as an empowered, self-organized team, you may be able to use Kanban.

Sprints

Kanban doesn't utilize timeboxes or deadlines (i.e., sprints), so it takes a disciplined team that can work at a sustainable pace to consistently deliver. The team must be measuring throughput and be good at breaking their work down into smaller pieces, which helps them estimate when larger projects will be finished. If your team generally works on smaller pieces of work, they're likely a good candidate for Kanban. If your team works on projects that must be designed, broken down, and would benefit from interim deadlines, Scrum is for you.

Artifacts

Kanban has one artifact—the Kanban Board. Your backlog lives in your "to-do" column, and it should be prioritized like a product backlog or sprint backlog in Scrum. If your team's backlog isn't too unwieldy, then Kanban may be a good choice for you! Teams may have a separate backlog for work they want to get to immediately—a "this week" column, for example. I have a Kanban board in Trello for my personal to-do list, and I handle my work by pulling the items I expect to get done that week into a separate column and concentrating on those. Each Friday, I review the to-do list and pull items into the "this week" column. Those are the items that I'll work on starting Monday.

Where Does the Work Come From?

If work comes to the team from many different places, you may be a good candidate for Kanban. For example, a UX or DevOps team that gets requests from many product teams will generally thrive in a Kanban environment. A public Kanban board has the added benefit of letting everyone know where their work is in the queue. IT, customer success, and support teams are usually a great fit for Kanban as well. Even marketing departments tend to love Kanban. However, if your team is working on one large project, Scrum may be a better fit.

Metrics

If you're using Kanban, tracking metrics like throughput, WIP, and cycle time is important. Kanban doesn't have built-in meetings or retrospectives, so to ensure the team is performing at its best, someone must keep an eye on those metrics. Scrum metrics like velocity and burndown should, of course, be tracked as well. Many tools, like JIRA, provide visuals for both Scrum and Kanban metrics. So, if you're a chart geek, check out what your toolset already has built in.

Events and Culture

If your team enjoys its low-key, low-process culture but needs a place to stay organized, start with Kanban. Kanban is lightweight and is usually a palatable way to embrace an agile culture. Teams can always evolve their processes once they see what Kanban can do for them. I've had many teams start on Kanban and then step into doing sprints or holding a single meeting each week. Sometimes they go all the way to adopting Scrum, but more often they find the right mix of "Scrumban" for them.

Kanban: Reality

Unfortunately, many teams aren't disciplined enough when it comes to Kanban. They don't track work in progress (WIP) limits or throughput, or they often overrun their WIP limits.

Scrumban

As I mentioned, a construct called Scrumban is a combination of Scrum and Kanban,[5] and it can actually be quite a good framework. Teams take elements from both Scrum and Kanban to make them run more effectively. Scrumban comes in many different flavors. I've seen teams start with Kanban, then add a daily stand-up for a while. They may add a planning meeting or two, and then stick with that process. Some teams even run sprints with Kanban, so they have deadlines for certain pieces of work. I recommend you start with one of these combinations, and then keep what works and continuously improve what doesn't.

Kanban: Survival Tips

1. Kanban Board - If your team isn't using a visual that shows all their work in progress yet, it's easy to create a Kanban board. Most teams use a tool like JIRA, Git Tracking, etc. for tickets. We're too distributed in most workplaces to not do so. All these tools have a Kanban view; they may call it something different, but a quick Google search should help you find what you need. Set up that view for your existing tickets—it only takes five to ten minutes to create it. Then, you can see how your team likes the view.

2. WIP - Definitely set WIP limits for your teams, as they force you to prioritize, get work done, and acknowledge when work is blocked. Also, it feels damn good to get things done! Start with a WIP limit of 1.5 to two times the number of people on your team and adjust if needed.

3. Meetings - Even if you're doing pure Kanban, have at least one short stand-up per week to connect with your team and get organized. Have whoever is the best time-keeper timebox the meeting. You'll be surprised at how useful these touchpoints can be.

4. No Meetings - If your team genuinely hates meetings, you can skip them. Someone like a team lead must take charge of organizing the backlog and making sure work is moving, however. You can use slack or a similar tool to keep everyone

updated on what you're working on and when you're blocked.

5. Count Tickets - Every two weeks or each month, count how many tickets your team completed. That number is your throughput rate, and it gives you a good idea for how quickly you'll get to future work. Someone important will always ask how soon you'll get to something, and this way you can give them an educated guess without spending hours on estimating.

6. Start with Kanban - If you're an engineer or entrepreneur who wants to try Scrum, but your team is pushing back, start with Kanban. Layer on a few concepts from Scrum if you'd like. You'll get 80 percent of the benefits with a lot less process.

Chapter 7: More Key Concepts and Survival Tips for Engineers

T his chapter covers a handful of topics that fit in any agile framework and that should give you an extra boost of knowledge as you begin your career. You may come across these elements in your first few jobs, or you may decide to try one or more of these techniques after you read about them. Some are easy to try on your own, while others require some team coordination.

Pair and Mob Programming

Pair and Mob Programming: Key Concepts

Pair Programming

One practice that is useful (and worth trying today, in my opinion) is called pair programming.[1] Pair programming is when two developers sit down together and work on the same code—one typing and the other dictating a solution. Every so often, the pair switches roles and continues. They often stop and discuss various solutions in real-time, and then they continue typing once they reach a consensus. When it comes to coding, two heads really are better than one!

Why?

Problems can often be solved in many ways, and with two developers working together, you double the number of ideas. Pairs almost always land on a solution that is more efficient and elegant because your partner is a built-in code reviewer. When bugs arrive later, both of you are familiar with the problem and able to solve issues.

From a teaming perspective, pair programming builds camaraderie and can be more fun. Developers learn each other's strengths and weaknesses and begin to compensate for them. Backup is built in for everything on which a pair works together, which allows developers to more easily take vacation or switch roles.

Would you want to pair on everything all the time? Most teams don't, but some businesses, like Menlo, Inc. have become famous for regular pairing.[2] They always pair program and rotate pairs every week. Most companies still have work that makes sense to do alone, but training, scaling, and solving tough problems can all be more efficient and effective in pairs.

Mob Programming

Mob programming is like pair programming, except the whole team participates.[3] One team member will type, while the others discuss and dictate code. The team switches typists every fifteen to thirty minutes.

With mob programming, you get many of the pair programming benefits, but you add even more ideas to the mix. Mob programming isn't for every problem, though; it works best when teams are working on particularly key, new, or challenging solutions. It also works well for new teams just starting to get to know one another. Some teams prefer to have a standing two- or three-hour "mobbing" meeting on the team's calendar. They can usually find something to work on during that time; if not, they cancel. The time together helps build camaraderie, and teammates learn a lot together. Plus, best practices and norms are more easily shared across the team.

Pair and Mob Programming: Reality

Many managers are worried that developer throughput will be cut in half with pairing, but in reality, that isn't the case.[4] A joint study by Laurie Williams (North Carolina State University), Robert Kessler (University of Utah), Ward Cunningham, and Ron Jeffries (one of the original signers of the *Agile Manifesto*) found that solving a problem with pair programming took only 15 percent more time, while bugs decreased by 15 percent.[5] Tech debt may also be reduced, as each developer brings their top ideas and the team picks the best one. That way, the code tends to need less refactoring. Time spent seems like a pretty solid trade-off for all the benefits mentioned above!

Pair and Mob Programming: Survival Tips

1. Try it! - Grab a friend and a tough problem and see whether pair or mob programming works well for you. Switch places every half-hour to an hour. Chances are, you've already worked on projects like this in college, so you may already know how much fun it is to pair on problems. The fun doesn't have to stop when you've graduated! However, I know pairing isn't always for everyone. Stopping senior engineers to ask questions can be intimidating, but communicating what you're doing or pointing things out in a helpful way is a worthwhile skill to build.[6]

2. Remote Teams - You can pair in person or remotely by sharing your screen and getting on a video call. Pair or mob programming remotely is as effective as it is in person.

3. Concentration - Concentrating while pairing can be difficult, which is one of its downsides. However, many engineers find that the benefits more than offset that concern. Having another person to keep track of loose ends and improve ideas is worth it.[7] If you need to think deeply about a problem sometimes and feel you want to take a pass alone first, do that! Then, have another engineer review your code later.

Branching and Merging Code

Branching and Merging Code: Key Concepts

When working on a team, code management becomes an essential part of a developer's job. Software-as-a-service (SaaS) products always have a copy of the code running that customers are using. We call this production. We try to keep production as stable as possible, so when developers work on code, they work on a copy of it. Today, two main development models are available for delivering quality software: Gitflow and trunk-based deployments.[8] Gitflow is somewhat based on a third workflow: feature branching. Each company may branch differently, so I'll go over these common workflows and their pros and cons.

Branches

Branches are a copy of the code managed in a version control system (VCS).[9] Various version control tools are available, the most popular being Git. Subversion has also maintained a following, but for this section's purposes, let's assume we're using Git. Code gets changed in a side branch, committed back to a branch, and reviewed. Eventually, it is committed to the code's production copy. How this happens depends on the workflow your team is using.

Merging Code and Conflicts

When code is committed by multiple people to a shared branch for the same file or line of code, a merge conflict may occur. The second person to commit will receive a merge conflict warning,[10] which means someone needs to look at the changes and decide how to incorporate both or decide which one should override the other. Merge conflicts like this are common, especially on larger teams.

Feature Branching and Gitflow

A developer creates a feature branch to work on an entire feature for the codebase. She won't commit her changes to a shared branch until she's completed her part of the feature. Feature branches tend to be long-lived branches; when they're finally committed to a shared branch, they may create several merge conflicts.[11]

Gitflow is similar in that it also has longer-lived branches and larger commits. It has a main (production) branch and a development branch. Feature branches are built from and committed back to the development branch, and that branch only gets merged back into main once the entire feature or project is completed. Individual developers may work on the feature branches or on an individual branch that gets merged back into the feature branches.

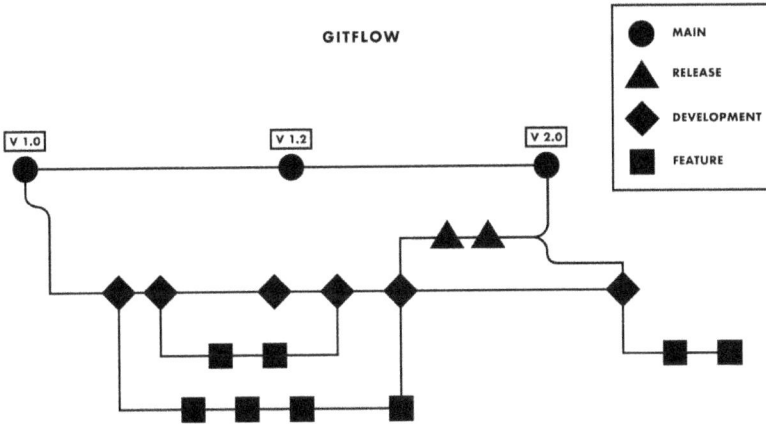

GITFLOW

●	MAIN
▲	RELEASE
◆	DEVELOPMENT
■	FEATURE

Release branches may be created from the development branch. Testers can test on any branch, but a full suite of test cases run on a release branch before it's merged back to main. When complete, that release is tagged with a version number. You'll notice that the full history of all code changes occurs at the development branch level, whereas main tends to be an abridged version of all changes made.[12] Finally, hot fixes or patches (i.e., changes made for high-priority bugs) have their own branch off of main. They're merged back into the main and the development branches. Let's look at a very simple example using Gitflow.

I'm a developer, and I'd like to change a button color from pink to blue for the next release. So, I create a feature branch off of development, make my change, and commit it back to the development branch. Best practice says I would tag a teammate to look at the code and make changes if I agree with their suggestions. Testers on my team may now test my color change. When everyone's changes are finished and committed, we create a release branch based on the development branch and testers will test the entire release. When it's ready, the code is committed to main and customers will see my new blue buttons.

Only developers with certain permissions generally commit code to production. They might be a team lead or DevOps engineer—that's the happy path. If a merge conflict arises, one or both of the developers involved handles it and changes are committed to the development branch.

You can't avoid merge conflicts unless you have a team of one. The more code that's changed, the more opportunities arise for merge conflicts. The best way to avoid conflicts is to make small changes and commit them often. Sound familiar? That's what we've been talking about doing with agile software development this whole time! Small, incremental changes make our lives much easier!

Trunk-Based Deployments

For trunk-based deployments, all developers have access to main. Developers merge smaller changes more often to the main branch, or "trunk." This way, far fewer changes occur in each commit, so testing overhead decreases and the chances of merge conflicts or new production issues go way down. Trunk-based deployments keep releases moving, which allows teams to deploy code more often. With quicker releases, code reviews, test automation, and feature flags (see the next section) become more important.[13]

TRUNK-BASED DEPLOYMENTS

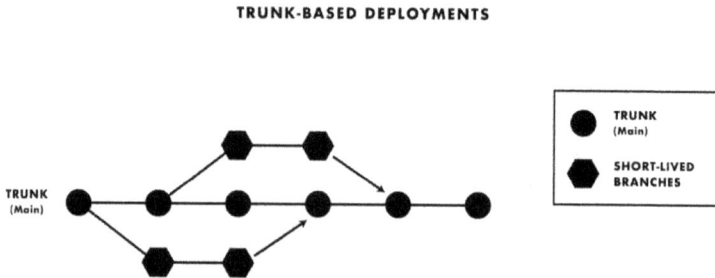

Branching and Merging Code: Reality

Merging code can get gnarly. Some people merge often, and some people hold onto their work for weeks before merging it, so when they do, merge conflicts arise. Sometimes, entire teams can spend days or weeks untangling the mess.

Some teams insist that their work can't be tested until a large portion is completed, and sometimes that's true. However, too much interconnecting code leads to large code merges, causing a mess of conflicts. Even if this problem doesn't happen, holding onto code for too long is simply not good for the product. Dropping a bunch of code at once makes testing difficult, can bottleneck other projects, and doesn't lend itself well to agile customer feedback.[14]

Long releases tend to create a similar pattern. More code goes into each release, which means it's harder to test and opens up more chances for bugs to get into the production code a customer uses. Also, you'll have fewer chances to receive valuable customer feedback.

Branching and Merging Code: Survival Tips

1. Commit Often - Commit and merge code as often as possible—ideally once a day or multiple times per day if you're actively coding. Sarah Murphy, one of our engineering managers at Singlewire, suggests you merge code, "When you feel like you've done meaningful work and your code compiles and builds fine." If you're working on a tough problem all day, you may not have any code to commit, which is okay. Committing often will help you avoid dealing with huge merge conflicts.

2. Short Releases - The less code that goes out in a release, the easier it is to manage. How short is short? Follow the same rule of thumb as for sprint length—less than four weeks is ideal. As a new developer, you may not have a say over how long your releases are, but you can at least ask about it when you're interviewing for jobs.[15]

3. Code Reviews - Commit your code, and then ask a team member to review it and offer suggestions for ways to improve it. This is a great way to learn from more senior engineers and make sure the quality of code shipped is high. Sarah Murphy also suggests: "Don't be afraid to ask to review senior engineers' code. You can learn a lot from reading their code and trying to understand the way they do things. Some will even be receptive to you asking questions or having discussions about the code, so you can deepen your understanding." Also, always review your own code before submitting it to others for review. Your time and your teammates' time are valuable.

4. Breaking Things - Another great insight from Sarah: Don't be afraid to break things! She says, "You might think you can blow

up your team's workflow by doing something weird with Git. You'd have to try pretty hard, though, and people can magic things back into order pretty quickly. Commit to the wrong branch? Usually pretty fixable (bad, but fixable). Named something wrong? Fixable. Git is meant as a history, and you're not going to be able to blow all of that away unless you're trying."[16]

Feature Flagging

Feature Flagging: Key Concepts

One of the safest ways to release new features is to do it behind a feature flag. A feature flag is a toggle that allows you to turn code on or off in production. It's essentially an "if statement" that wraps around an entire feature. If a feature flag is set to "true" in a configuration file, then the feature runs in production. If it's not, it doesn't run. You can release code little by little behind a feature flag and then turn on the feature once the entire thing is in production.[17] You can also use feature flags on multiple levels. You can set them on a customer-by-customer basis, allowing you to test new features on a subset of your user base, and then deploy to all when you're satisfied. This type of release is called a "canary release" and can be useful when rolling out large changes.[18]

Feature Flagging: Reality

While feature flagging is a best practice, not every organization uses it. Sometimes, code is not organized in a way that makes feature flagging easy.

Feature Flagging: Survival Tips

1. Feature Flag Everything - Feature flagging any major feature gives you flexibility. If a feature gets into production and bugs start streaming in (or worse, it takes down the whole product), all you need to do is turn it off while you investigate. Rushing around to release a fix is therefore unnecessary. When you're

out looking for jobs, ask about whether the company is using feature flagging. The practice is a sign of maturity that will lessen the number of high-priority issues you'll need to handle "right now."

2.   Tooling - Tools are available that will help you feature flag your product.[19] If your company is looking to start using feature flags, some may be worth investigating. However, you don't need tools to get started—you can build feature flags into your product with if statements and configuration files.

Test-Driven Development (TDD)

Test-Driven Development: Key Concepts

Along with the code, developers write automated unit tests, which test the code sections they've written. The tests are run each night, each release, or even every hour. They tell you whether previously-working code broke since the tests last ran. On most teams, engineers write unit tests after their code.

Test-driven development (TDD) flips this practice on its head.[20] Using TDD, developers write automated test cases first. They run the test cases often and write the code until the test cases pass.

How can this method be useful? Won't you have a bunch of failing test cases in the system at any given moment? Yes. Yes you will, and that's okay. Writing the test cases first forces you to think through the code from a tester's perspective. You'll think through the happy path and the edge cases before writing one line of the actual code, which tends to help engineers think more holistically about their code. Quality work is baked in.[21]

Another benefit is that sometimes we all run out of time and cut corners on our test cases; that's easy to do. No one notices if a test case is missing, but everyone notices if a test case isn't passing. TDD ensures corners don't get cut on quality by ensuring the test cases are there before the code appears.

Test-Driven Development: Reality

Organizations tend to embrace the practice of test-driven development . . . or they don't. Some companies struggle to build test cases that cover their entire code base. A lot of legacy code was written before unit testing was commonplace, so if you have 80 percent test coverage for any given feature, you're doing pretty well.

Test-Driven Development: Survival Tips

1. Try it locally - If your organization doesn't use TDD yet, try it out in your local environment to see how you like it. That way, you won't break the suite of automated test cases already running. No one will bother you about your broken tests.

2. Communicate - Let people know you're using TDD. One of the easiest ways to get your team on board is to show them how well it works.

DevOps

DevOps: Key Concepts

DevOps is a philosophy, culture, and role that is almost never covered in university courses. But DevOps is an important part of a high-performing agile team. DevOps, like agile software development, is a culture and a mindset.

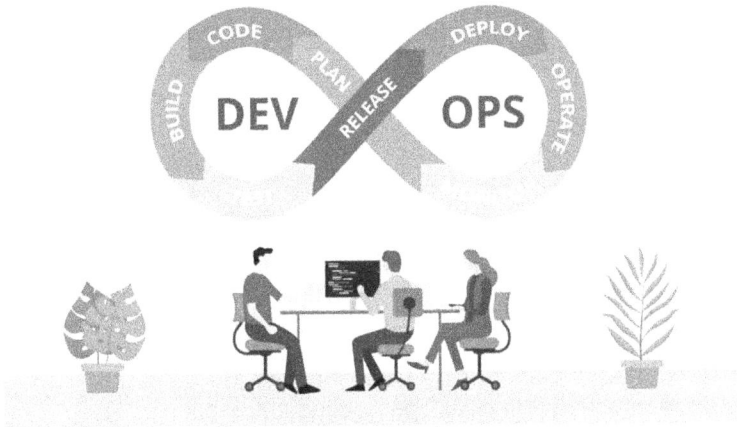

Image courtesy of vecteezy.com

DevOps is the merger of operations and development. Operations make sure your production products are always up and running, and development automates tooling and infrastructure to make everyone more efficient. Someone in a DevOps role does a variety of things on any given day. They might spend time setting up production monitoring for a new feature, responding to production issues, releasing to production, setting up a test environment, writing a script to automate part of their job, testing out the latest and greatest tool, or working on a larger project to completely upgrade a major part of the department's infrastructure. DevOps engineers do these things and so much more.

A high-performing DevOps team automates nearly everything they do. For instance, environment set-up is self-service. A tester can press a button and have a new testing environment in a matter of minutes. The release process is automated and happens on a weekly, daily, or even hourly basis. Testing is automated and robust monitoring is in place.[22] It takes a while to get to this ideal state, but it's well worth the time and effort. Does the DevOps team get disbanded at this point? No way! On the contrary, they get to work on the super fun stuff like exploring new tools, optimizing what they've built, and of course working with teams to release cool new features!

DevOps: Reality

DevOps as a concept came about several years after the *Agile Manifesto* was signed. Most DevOps organizations are not as polished as the high-performing example above; in fact, some can be quite immature. A lot of companies out there will try to sell teams tools that will make them "DevOps," but no tool can do that because it's a culture, not a tool. Some organizations don't have a DevOps team. Instead, they have their agile teams monitor and attend to production for the code they've created, which can be very effective too (although it's nice to have a team focused on developer tooling automation).

DevOps: Survival Tips

1. Automate Everything - Even if you're not on the DevOps team, if you need to do something more than twice, automate it.

2. DevOps as a Career - My friend Sara Willett runs an arm of the DevOps organization at MasterCard. She says the types of people who do well in DevOps are "lazy developers." In other words, they're the type of people who don't like to do the same thing more than twice, so they automate it. They're the developers who are interested in trying out new tools or languages, and they're good at figuring out how to integrate tools with other tools. DevOps engineers are the people who like to tinker and who like to do a variety of tasks; they don't mind context switching. If that sounds like you, DevOps may be an interesting career path for you! DevOps engineers are in demand right now—those with some experience can command high salaries. You can apply to DevOps roles right out of college and may find organizations willing to hire and train you. If not, another path into DevOps is to get a software development role at an organization with a DevOps team. Ask a member of that team to mentor you into a position. You can also go out and get an AWS or Azure certification to supplement your engineering degree, which may help you stand out during the DevOps interviewing process.

Agile at Scale

Agile at Scale: Key Concepts

You may have noticed that everything we've discussed so far revolves around a small, core team of less than ten people. What happens if we have a large organization with thirty, fifty, or hundreds of engineers building at scale? In that case, you take these basic building blocks we've discussed and add some layers of coordination, making sure to preserve the same agile principles throughout the system. Agile at scale has several flavors. For example, the Spotify model[23] and Scaled Agile Framework (SAFe)[24] are very similar models. Large-Scale Scrum (LeSS)[25] and Disciplined Agile Delivery (DAD)[26] are important frameworks too.

The only scaled agile model I like is the Spotify model, and it's not even a real agile framework. I've used it and the Scaled Agile Framework during my career, both with some degree of success. The Spotify model is a collection of practices that the teams at Spotify have found useful. It's very similar to Scaled Agile Framework, but with less process and fewer specialized roles. I will discuss the Spotify model here only to give you a taste of how agile at scale can work. I recommend you learn about one or two of the other frameworks if you're working in a large organization. The "Notes" section at the back of this book has links to some resources.

Henrik Kniberg (one of my favorite agilists) and Anders Ivarsson published a paper in 2012 called *Tribes, Squads, Chapters, and Guilds,*[27] which explains the team structure at Spotify.

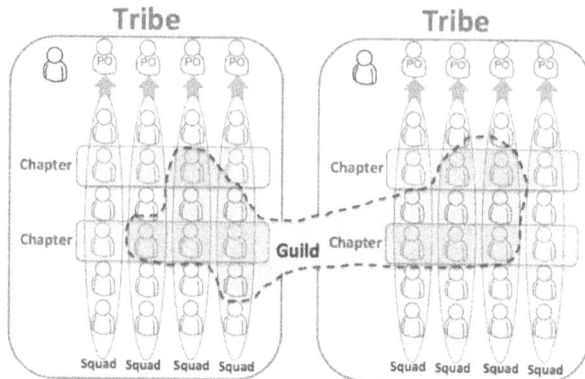

Spotify Team Model

Squads are agile teams just like we've been discussing so far in this book. They're small, self-organized, dedicated, cross-functional, and empowered.

Tribes (also known as "programs") are a group of teams that work on interconnected features. For example, when I worked at Opower, we had a tribe for each of our products. We also had a platform tribe which maintained the code that interacted with all our products. Our largest product was the Home Energy Report (HER)—a report we sent to customers showing them how much energy they'd saved (or hadn't) compared to their neighbors. A tribe was associated with our HER product, and within that tribe, we had six squads working on different features for the Home Energy Reports. All these teams worked on the same three-week sprint cycle and delivered on specific, coordinated dates. Each tribe had an engineering director and product director, and they organized the tribe across teams. Tribes also had an agile program manager who acted as Scrum Master for the individual teams. They ran sprint planning and other ad hoc ceremonies as well, whenever the tribe needed to sync up.

Squads within tribes could create silos (teams that don't talk regularly). To break down those silos, Spotify created two concepts they call chapters and guilds.

Chapters are self-organized groups of people with the same job role. They meet on a regular basis to share best practices and talk about what they're doing. Testers, for example, may form a chapter to share learnings and best practices.

Guilds are self-organized groups that share an interest in a particular topic. Engineers who write in Java, for example, may form a guild to discuss how other teams are using Java.

Both groups are led by volunteers. The riskiest part of having chapters and guilds is finding enthusiastic leaders who will make sure they meet on a regular basis and discuss important topics.

Those are more or less the basics of what the Spotify structure entails. But Spotify also leans on its culture to ensure agile at scale runs smoothly. Spotify has published a lot of information about its culture. They have a two-part video series that's worth viewing (Part I[28] and Part 2[29]).

Spotify's key driving force is autonomy. They aim to make each squad as autonomous as possible, and their mission is "loosely-coupled, but tightly-aligned." Each team decides what to build and how to build it. Their architecture is as decoupled (i.e., free of dependencies) as possible, so squads can release new features when they're ready. Squads work in whatever way works for them (Scrum, Kanban, etc.), but they have a strong culture of sharing. When a practice gains enough momentum, other squads adopt it. Teams run health checks to track their maturity and happiness,[30] which is their way of driving continuous improvement.

Engineers at Spotify pride themselves on their culture of respect, trust, and helpfulness. In the Part I video, Kniberg says, "Agile at scale requires trust at scale." Any successful organization needs this trust to work effectively at any size. Part of that confidence is trusting people to fail and learn from it. Failure is not seen as negative; it's seen as a learning opportunity. Squads hold post-mortems and retrospectives on a regular basis to drive continuous improvement.

Squads at Spotify aim to avoid waste at all times and experiment to get rid of it whenever possible. They strive for "just enough" process and to adopt the easiest practices that work. For example, many elaborate tools and practices are available for dealing with dependencies, but Spotify uses a simple-yet-effective spreadsheet to keep track of cross-squad dependencies. Of course, they avoid dependencies as much as possible, as dependencies cause projects to take longer.

Spotify's product development process is based on lean start-up principles. They know the riskiest thing about product development is building the wrong thing, so they build a minimum viable product. They then deploy it, learn, and iterate, exactly like we discussed in "Part I: Discovery."

Innovation is also important. At Spotify, they care more about innovation than predictability, but they do use agile planning techniques if they need to have a date for, say, marketing purposes. Employees are encouraged to spend about 10 percent of their time on "hack time," which means they work on innovative projects that aren't tied to their regular work. They also hold hack-a-thons in which employee ideas are actually deployed to customers. Then, they look at the data from what they've released to find places to improve their features.

Spotify's model is simple and it works, making it worth a close look for large teams trying to release quickly. A few articles have been published about Spotify not *really* using its own model, and if that's true, it's too bad. We did use many of these practices successfully at Opower, which is where my strong endorsement of the process originates.

Agile at Scale: Reality

I don't think anyone has the magic formula that allows large businesses to move as quickly as smaller organizations. All these frameworks try their hardest, but if you want to work somewhere that ships fast, your best bet is a small organization.

However, learning more about agile at scale if you work in a large organization is worthwhile. Look-up content online about Scaled Agile Framework (SAFe), Large Scale Scrum (LeSS), or Disciplined Agile Delivery (DAD) to find out more.

Agile at Scale: Survival Tips

1. Size Matters - I'm grateful to have worked in small and large organizations over the years because those experiences have helped me figure out which I like better. You can explore for yourself too. Do multiple internships at different companies while you're still in school; that way, you can make an informed decision when you graduate. All companies have their pros and cons. Large organizations tend to have tried and true processes, plus opportunities to work on different products, but your sphere of influence over a product will be smaller. Some offer higher pay as you climb the ladder, and employee stock programs may be available. Many have better name recognition for your résumé. They also have more bureaucracy and process in general. It can be harder to grow your career, as you have to jump through more hoops to get large promotions along the way. On the other hand, smaller organizations tend to move faster. You can try out different positions, use newer technologies, likely be responsible for knowing a broader part of the tech stack, and have a bigger influence on decisions. Their pay scale depends on how well-

funded the organization or start-up is. You may get a lot of equity, but have to wait until the company is sold or goes public before you can cash in. Sometimes, these companies are less stable and disappear.

2.   The Big Picture - When you're at a large company, it's easy for the big picture to get lost. Why are you building this product? Who's it for? How is it improving their lives? Understanding your product's big picture is very important. What is the product vision? How do the various teams and departments contribute to that vision? How does your team contribute to it? The more you understand about why you're building the product, the better you'll be able to do your job.

Business Agility

Business Agility: Key Concepts

Agility is not just for software developers. The most effective companies embrace agile principles in all departments, from the C-Suite and budgeting practices to marketing, sales, and of course their agile software teams. The agile principles and concepts in this book plug nicely into other areas of the business.

To learn more about business agility, visit the Business Agility Institute website at https://businessagility.institute/. This organization was founded by the talented Evan Leyborn and Ahmed Sidky. Or, go to one of their wonderful conferences held around the world throughout the year.[31]

Business Agility: Reality

Few tech companies embrace agility throughout their organization. Making an entire organization more agile takes a forward-thinking executive team. A new engineer likely won't be able to dictate how marketing runs, but you can look for signs that a company is well-run when you're hired. If you're starting your own company, embrace the agile principles wherever you can.

I could talk at length about business agility. All I'll say in this book, though, is that when you're looking for jobs, evidence that business agility exists is a very good sign.

Business Agility: Survival Tips

1. Signs a Company Embraces Business Agility

    a. The CEO is accessible and clearly articulates the company strategy. She encourages all departments to continuously improve the way they work.

    b. Company-wide budgeting and planning are performed quarterly on a rolling basis, not once a year.

    c. Customer interviews are conducted often and learnings are shared throughout the company.

    d. Software teams are using agile development techniques. Your interviewers speak enthusiastically about their agile culture.

    e. The company makes changes based on what existing customers are saying. Sales and prospective customers are one input to building the roadmap, not the only input.

    f. Marketing and/or customer success teams get excited when you explain to them what Kanban is. Maybe they already use it?

    g. Innovation is baked into the culture. Ideas are tested rapidly, and the good ones find their way onto the official roadmap.

    h. Everyone across the company knows what the current mission and product strategy is.

2. Start with Kanban - Kanban is applicable in so many areas outside of software development. It's easy to implement and boosts efficiency immediately. I find that marketing, customer success, and IT teams in particular love Kanban. National

Geographic's marketing team adopted it during our agile transformation. If you're ever in a position tasked with increasing a team's productivity and don't know where to start, try Kanban.

Conclusion

There you have it: agile software development in a nutshell! The material in this book covers the essentials I wish I'd known when I graduated with my Computer Science degree. So much of success in software is about building the right product and collaborating effectively with your teammates, and life also tends to be better when you enjoy the people you work with and the way you work. The agile principles contribute not only to the product's success, but to the happiness of the people who work on it (that's us!). Don't ever work anywhere if you're not having fun—always make sure you're excited about the purpose behind what you're building. If you use what you've learned in this book and enjoy what you do, you'll have a very fulfilling career.

Agility Everywhere

The principles in this book don't only work for agile software teams; they're useful in all areas of life. Many people have found success applying them to their personal lives as well. For example, I use a Kanban board to keep track of my personal to-dos. Some enterprising parents have instituted daily stand-ups for their families. The sky's the limit.

Wherever you go with this knowledge, I hope you find it useful as you start your careers in technology. It's a great field, filled with possibilities. Enjoy your career and the time you spend working on agile teams!

Where to Find Me

I hope you've found something useful in this book; if so, I would love to hear from you!

I blog at amberrfield.com. You can subscribe or reach out to me there using the Contact link. Of course, I also teach at the University of

Wisconsin-Madison. If you happen to be a student there, take my course. If you do, you'll have a giant head-start having read this book in the first place. If you work for a company that would like to explore becoming a partner for UW's Capstone class, you can find my contact info here: https://www.cs.wisc.edu/staff/field-amber/.

Finally, if you'd like some help trying self-selection, reach out to me at my contact link as well. I love helping organizations try out self-selection.

Acknowledgements

I thought this may be the toughest section of the book to write, but once I got started, the names and gratitude flowed freely. I only hope that I haven't forgotten anyone important here!

Thank you to my family, Graham, Alison, Maddie, Ceilia, and Theo, who have dealt with numerous hours of me typing away at home on the book. My parents, Kathleen & Tim Field, who gave me plenty of opportunities to write as a kid. To my oldest daughter, Alison King (she was 12 at the time of publication), who not only did the cover art for my book, but came up with the concept after just a few minutes of hearing me drone on about agile software development. My editor, May-Zhee Lim, whose insights were incredibly on point. Thanks to Kathryn Palmer who proofread the book and got rid of all of my silly grammar mistakes. Thanks to my graphic designer, Emily Varone, who just exudes joy and professionalism at every turn, and Win Treese, who recommended her. Thank you to Carin at GetCovers Design for coordinating the design team to quickly create an engaging book cover.

To the numerous writers, entrepreneurs, and go-getters in Writing in Community 3 and Brainstorm Road who read my very bad first drafts every day and helped me turn them into something much better: Win Tresse (again!), Susan Walter, Joyce Sullivan, Raj Nagappan, Lori Rodriguez, Maggie Croushore, Davender Gupta, Jessica Zou, Katharina Tolle, Terri & Bill Tomoff, Ashley Cirilli, Jacquie Clarke, Darma Singam, Devorah Graeser, Heather Button, Diane Wyzga, Roni Vayre, Jim Ward, Doug May, Patricia Bedard, Libby J, Julie Rains, Marianne van den Broek, Ciela Hartanov, George Stenitzer, Melissa Kalinowski, Anne Berg, Jennifer Soos, Steven Price, Janneke Ritchie, Jane Hasenmueller, Andrew Whitehouse, Kevin Liebenberg, Melissa Balmer, Phaedra Romney, Luis Iturriaga, Tom Nealley, Kevin Gilds, Alex Keerie, Katy Dalgleish, and Ken L. And to Seth Godin, Kristin Hatcher, and Margo Aaron who sponsored and moderated these vastly important groups!

To my beta readers who gave me wonderful input and encouragement: Jen Mann, Clayton Custer (https://www.edu-reality.com/), John Halberstadt (https://lithespeed.com/), Lauryn Branham, Aryan Adhlakja, and my entire CS639 Capstone Class, Fall 2022.

Thank you to my co-workers who saved a couple sections in Chapter 7 and provided great Survival Tips: Sarah Murphy, Charlie Ciccia, and James Elliott. And to Mary Jo Trapani who gave me feedback and great reference material from the book she edited.

Thanks to Remzi Arpaci-Dusseau, Chair of the Computer Sciences Department at the University of Wisconsin-Madison, who made the Capstone class a reality and gave me the tools I needed to turn it into a successful course.

Thanks also to my former manager, Larry Mcguire, and Singlewire's CEO, Paul Shain, who have been supportive of my career and of me teaching the class and writing this book in addition to working full-time at Singlewire.

To a few key players and friends throughout my career that have taught me an extraordinary amount about agility, collaboration, and life: Sanjiv Augustine (https://lithespeed.com/), Itamar Goldminz, Maureen Kearns, Terri Lundak, Sara Willett, and Joey Spooner. Thank you for being a part of this journey!

And finally to the co-authors of my first book, The Badass Sisterhood Anthology, who taught me all about the publishing process and how a group of women together can do whatever they'd like: Jackie Alcalde Marr, Katherine Cartwright, Wendy Coad, Maggie Croushore (again!), Micki Jean LaVres, Ellen Newman, Laurie Riedman, Cecilia Roddy, Terri Tomoff (again!), and Susan Walter (again!).

The process of writing this book has been fulfilling, enlightening, difficult, and, to steal a phrase from my first book, just plain badass all at the same time.

Updates & Suggestions

Just like the agile concepts I discussed in the previous chapters, this book itself is one iteration of many and I'd like your help to continuously improve it. If you find any typos or bigger errors that you'd like to see corrected in a future edition, please reach out to me. You'll find my contact information here: https://www.cs.wisc.edu/staff/field-amber/. As a thank you, I will include your name in a future book release!

Notes

Introduction
1. Rapoza, K. (2011, July 15). How to Survive the Death of Print Media.   Forbes. https://www.forbes.com/sites/kenrapoza/2011/07/15/how-to-survive-the-death-of-print-media/?sh=334fe2f33606
2. Dang, S. (2018, July 27). Fox and Disney shareholders approve deal   for   entertainment   assets.   Reuters. https://www.reuters.com/article/us-fox-m-a/fox-and-disney-shareholders-approve-deal-for-entertainment-assets-idUSKBN1KH1SD
3. Ries, E. (2011). The Lean Startup: How Today's Entrepreneurs Use Continuous Innovation to Create Radically Successful Businesses (pp 143). Crown.
4. Digital.ai. (2022, December 7). 16th Annual State of Agile Report. Resources | Digital.ai. Retrieved May 8, 2023, from https://info.digital.ai/rs/981-LQX-968/images/AR-SA-2022-16th-Annual-State-Of-Agile-Report.pdf
5. Digital.ai. (2022, December 7). 16th Annual State of Agile Report. Resources | Digital.ai. Retrieved May 8, 2023, from https://info.digital.ai/rs/981-LQX-968/images/AR-SA-2022-16th-Annual-State-Of-Agile-Report.pdf

Chapter 1
1. Agile Alliance. (2023). What is Agile? | Agile 101. Agile Alliance. Retrieved   April   7,   2023,   from https://www.agilealliance.org/agile101/
2. C-Suite Spotlight. (2022, September 22). Eric Yuan: Zooming in on One of Tech's Most Unlikely Success Stories. C-Suite Spotlight.   https://csuitespotlight.com/2022/09/22/eric-yuan-zooming-in-on-one-of-techs-most-unlikely-success-stories/#:~:text=Yuan%2C%20who%20served%20as%20corporate's,eventually%20called%20Zoom%20Video%20Communications.

3. Zoom Video Communications Inc. (2023). Zoom Releases by Date. support.zoom.us. Retrieved April 07, 2023, from https://support.zoom.us/hc/en-us/sections/360008531132-Zoom-Releases-by-Date

4. Mendoza, N. (2021, March 31). Zoom zips ahead of Google Meet, Microsoft Teams and Skype in one ranking. TechRepublic. https://www.techrepublic.com/article/zoom-zips-ahead-of-google-meet-microsoft-teams-and-skype-in-one-ranking/

5. Royce, W. W. (1970). Managing the Development of Large Software Systems. Proceedings of IEEE WESCON, (26), 328-388. https://www-scf.usc.edu/~csci201/lectures/Lecture11/royce1970.pdf

6. Scrum Institute. (n.d.). What Makes Waterfall Software Development Model Fail in Many Ways? ScrumInstitute.org. Retrieved 04 07, 2023, from https://www.scrum-institute.org/What_Makes_Waterfall_Fail_in_Many_Ways.php

7. Beck, K., & et al. (2001). The Agile Manifesto. agilemanifesto.org. Retrieved April 7, 2023, from agilemanifesto.org

8. Cialdini, R. B. (2007). Influence: The Psychology of Persuasion. HarperCollins.

9. Textor, C. (2022, February 11). Hong Kong: population breakdown by language. Statista. Retrieved April 7, 2023, from https://www.statista.com/statistics/329051/hong-kong-population-breakdown-by-language/

10. Sutherland, J., & Schwaber, K. (2020). Scrum Guides, 10. Scrum Guide. Retrieved April 7, 2023, from https://scrumguides.org/scrum-guide.html

11. Sutherland, J., & Schwaber, K. (2020). Scrum Guides, 5-6. Scrum Guide. Retrieved April 7, 2023, from https://scrumguides.org/scrum-guide.html

12. Duignan, B., & Vignaux, P. D. (2023, March 31). Occam's razor | Origin, Examples, & Facts | Britannica. Encyclopedia Britannica. Retrieved April 7, 2023, from https://www.britannica.com/topic/Occams-razor

13. McKeown, G. (2020). Essentialism: The Disciplined Pursuit of Less, 8. Crown.

14. Samarthyam, G., Suryanarayana, G., & Sharma, T. (2014). Refactoring for Software Design Smells: Managing Technical Debt, 3. Elsevier Science.

15. Sutherland, J., & Schwaber, K. (2020). Scrum Guides, 7-8. Scrum Guide. Retrieved April 7, 2023, from https://scrumguides.org/scrum-guide.html
16. Pang, A. S.-K. (2018). Rest: Why You Get More Done When You Work Less, 53-74. Basic Books.
17. So, why are typical work days 8 hours long? That has more to do with the Fair Labor Standards Act of 1938 (FLSA), than how we actually function. FLSA established overtime pay for people working over 44 hours (later reduced to 40 hours). This forced companies to re-examine their work schedules. Manufacturing companies realized if they could hold three shifts of 8 hours each per day. With those, they could run operations around the clock. That's the schedule that stuck. The 8-hour workday is not based on the science of how engineers become successful. It's based on trying to ensure factory workers didn't physically burn out. Horowitz, A., & Horowitz, A. (2021, October 27). More employees are looking to get their 40 hour work week lowered to 30 hours. : Planet Money. NPR. Retrieved April 7, 2023, from https://www.npr.org/2021/10/27/1049786108/nice-work-week-if-you-can-get-it
18. Measey, P. (Ed.). (2015). Agile Foundations: Principles, Practices and Frameworks. BCS Learning & Development Limited. Ch. 14
19. Business Agility Institute. (2023). Frequently Asked Questions. Business Agility Institute. Retrieved April 7, 2023, from https://businessagility.institute/faq#faq-01

Chapter 2
1. Sutherland, J., & Schwaber, K. (2020). Scrum Guides, 5. Scrum Guide. Retrieved April 7, 2023, from https://scrumguides.org/scrum-guide.html
2. Leopold, K., Hazel, C., & Magennis, T. (2020, June 3). The Impact of Dependencies -- and How to Manage Them. You Tube. Retrieved April 12, 2023, from https://www.youtube.com/watch?v=0Ixrx3Z9194
3. Mamoli, S., & Mole, D. (2015). Creating Great Teams: How Self-selection Lets People Excel. Pragmatic Bookshelf.
4. DeWyze, J. (2023, February 16). The 25 Percent Rule for Tackling Technical Debt (2023). Shopify Engineering. Retrieved April 12, 2023, from https://shopify.engineering/technical-debt-25-percent-rule

Chapter 3
1. Andreessen, M. (2007, June 25). Part 4: The Only Thing That Matters. Pmarchive. Retrieved April 13, 2023, from https://pmarchive.com/guide_to_startups_part4.html
2. Scientific method | Definition, Steps, & Application | Britannica. (n.d.). Encyclopedia Britannica. Retrieved April 13, 2023, from https://www.britannica.com/science/scientific-method
3. Ries, E. (2011). The Lean Startup: How Today's Entrepreneurs Use Continuous Innovation to Create Radically Successful Businesses. Crown.
4. Blank, S. (2010, April 29). Teaching Customer Development and the Lean Startup - Topological Homeomorphism. steveblank.com. Retrieved April 13, 2023, from https://steveblank.com/2010/04/29/teaching-customer-development-and-the-lean-startup-%E2%80%93-topological-homeomorphism/
5. Blake, S., & Womack, J. P. (n.d.). Understanding Lean Agile and the 5 Lean Principles. Easy Agile. Retrieved April 15, 2023, from https://www.easyagile.com/blog/lean-agile/
6. Ries, E. (2011). The Lean Startup: How Today's Entrepreneurs Use Continuous Innovation to Create Radically Successful Businesses, 9, 22. Crown.
7. Ries, E. (2011). The Lean Startup: How Today's Entrepreneurs Use Continuous Innovation to Create Radically Successful Businesses, 61-63. Crown.
8. Osterwalder, A. (2020). The Value Proposition Canvas. strategyzer.com. Retrieved April 15, 2023, from https://www.strategyzer.com/canvas/value-proposition-canvas
9. Ries, E. (2011). The Lean Startup: How Today's Entrepreneurs Use Continuous Innovation to Create Radically Successful Businesses, 57-58. Crown.
10. Launch Tomorrow. (2019, November 13). Why Your Riskiest Assumption is a Great Place to Start with Any New Product or Idea. Launch Tomorrow. https://www.launchtomorrow.com/2019/11/why-your-riskiest-assumption-is-a-great-place-to-start-with-any-new-product-or-idea/
11. Ries, E. (2011). The Lean Startup: How Today's Entrepreneurs Use Continuous Innovation to Create Radically Successful Businesses, 92-113. Crown.
12. Norman, D. (n.d.). What is Paper Prototyping? | IxDF. Interaction Design Foundation. Retrieved April 21, 2023, from

https://www.interaction-design.org/literature/topics/paper-prototyping

13. Hegedus, G. (2017, March 9). Mobile Application Design : Paper Prototype Video. YouTube. Retrieved April 21, 2023, from https://www.youtube.com/watch?v=y20E3qBmHpg

14. Stanford Design School. (2018, November 8). An Introduction to Design Thinking In One Hour (The Wallet Project). Stanford Design School. Retrieved April 15, 2023, from https://drive.google.com/file/d/1NNoG2RV0sJzW0TbECszgkpj Qq8e-vrDI/view

15. Babich, N. (2017, November 29). Prototyping 101: The Difference between Low-Fidelity and High-Fidelity Prototypes and When to Use Each | Adobe Blog. the Adobe Blog. Retrieved April 21, 2023, from https://blog.adobe.com/en/publish/2017/11/29/prototyping-difference-low-fidelity-high-fidelity-prototypes-use

16. Ferriss, T. (2009). The 4-Hour Workweek, 179-199. Harmony/Rodale.

17. Ries, E. (2011). The Lean Startup: How Today's Entrepreneurs Use Continuous Innovation to Create Radically Successful Businesses, 57-58. Crown.

18. Ries, E. (2011). The Lean Startup: How Today's Entrepreneurs Use Continuous Innovation to Create Radically Successful Businesses, 99-103. Crown.

19. Launch Tomorrow. (2019, November 13). Why Your Riskiest Assumption is a Great Place to Start with Any New Product or Idea. Launch Tomorrow. https://www.launchtomorrow.com/2019/11/why-your-riskiest-assumption-is-a-great-place-to-start-with-any-new-product-or-idea/

20. AirBNB. (2022, February 15). Airbnb fourth quarter and full-year 2021 financial results. news.airbnb.com. Retrieved April 15, 2023, from https://news.airbnb.com/airbnb-fourth-quarter-and-full-year-2021-financial-results/

21. Ries, E. (2011, October 19). How DropBox Started As A Minimal Viable Product. TechCrunch. https://techcrunch.com/2011/10/19/dropbox-minimal-viable-product/

22. Ries, E. (2011). The Lean Startup: How Today's Entrepreneurs Use Continuous Innovation to Create Radically Successful Businesses, 97-99. Crown.

23. Ries, E. (2011). The Lean Startup: How Today's Entrepreneurs Use Continuous Innovation to Create Radically Successful Businesses, 114-148. Crown.
24. Ries, E. (2011). The Lean Startup: How Today's Entrepreneurs Use Continuous Innovation to Create Radically Successful Businesses, 143. Crown.
25. Ries, E. (2011). The Lean Startup: How Today's Entrepreneurs Use Continuous Innovation to Create Radically Successful Businesses, 149-178. Crown.
26. Blank, S. (2013, May). Why the Lean Start-Up Changes Everything. Harvard Business Review. Retrieved May 9, 2023, from https://hbr.org/2013/05/why-the-lean-start-up-changes-everything
27. IDEO. (n.d.). Design Thinking Defined. IDEO Design Thinking. Retrieved April 15, 2023, from https://designthinking.ideo.com/
28. Stanford Design School. (n.d.). Field Notes — Stanford d.school. Stanford d.school. Retrieved April 15, 2023, from https://dschool.stanford.edu/field-notes
29. Ideo is a great example of a company putting Design Thinking into action, as they've been asked to reinvent everything from Target's shopping carts to Vertical Farming to Women's Breastpumps! IDEO. (n.d.). 11 Products Made Using Design Thinking. IDEO U. Retrieved April 15, 2023, from https://www.ideou.com/blogs/inspiration/11-products-made-using-design-thinking
30. Kowitz, B., Knapp, J., & Zeratsky, J. (2016). Sprint: How to Solve Big Problems and Test New Ideas in Just Five Days. Simon & Schuster.
31. Kowitz, B., Knapp, J., & Zeratsky, J. (2016). Sprint: How to Solve Big Problems and Test New Ideas in Just Five Days, 53-91. Simon & Schuster.
32. Kowitz, B., Knapp, J., & Zeratsky, J. (2016). Sprint: How to Solve Big Problems and Test New Ideas in Just Five Days, 95-123. Simon & Schuster.
33. Kowitz, B., Knapp, J., & Zeratsky, J. (2016). Sprint: How to Solve Big Problems and Test New Ideas in Just Five Days, 127-160. Simon & Schuster.
34. Kowitz, B., Knapp, J., & Zeratsky, J. (2016). Sprint: How to Solve Big Problems and Test New Ideas in Just Five Days, 165-190. Simon & Schuster.
35. Kowitz, B., Knapp, J., & Zeratsky, J. (2016). Sprint: How to Solve Big Problems and Test New Ideas in Just Five Days, 195-225. Simon & Schuster.

Chapter 4:
1. Cohn, M. (n.d.). User Stories and User Story Examples by Mike Cohn. Mountain Goat Software. Retrieved April 16, 2023, from https://www.mountaingoatsoftware.com/agile/user-stories
2. Patton, J., & Economy, P. (2014). User Story Mapping: Discover the Whole Story, Build the Right Product, 3 (P. Economy, Ed.). O'Reilly.
3. Patton, J., & Economy, P. (2014). User Story Mapping: Discover the Whole Story, Build the Right Product, 99-100 (P. Economy, Ed.). O'Reilly.
4. Patton, J., & Economy, P. (2014). User Story Mapping: Discover the Whole Story, Build the Right Product, 94 (P. Economy, Ed.). O'Reilly.
5. Patton, J., & Economy, P. (2014). User Story Mapping: Discover the Whole Story, Build the Right Product, 137-154 (P. Economy, Ed.). O'Reilly.
6. Cohn, M. (n.d.). Epics, Features and User Stories. Mountain Goat Software. Retrieved April 21, 2023, from https://www.mountaingoatsoftware.com/blog/stories-epics-and-themes
7. Cohn, M. (2004). User Stories Applied, 23-26. Addison-Wesley.
8. Sutherland, J., & Schwaber, K. (2020). Scrum Guides, 10-11. Scrum Guide. Retrieved April 7, 2023, from https://scrumguides.org/scrum-guide.html

Chapter 5:
1. State of Agile. (2022). 16th Annual State of Agile Report. stateofagile.com. Retrieved April 17, 2023, from https://info.digital.ai/rs/981-LQX-968/images/AR-SA-2022-16th-Annual-State-Of-Agile-Report.pdf
2. Sutherland, J., & Schwaber, K. (2020). Scrum Guides. Scrum Guide. Retrieved April 7, 2023, from https://scrumguides.org/scrum-guide.html
3. State of Agile. (2022). 16th Annual State of Agile Report. stateofagile.com. Retrieved April 17, 2023, from https://info.digital.ai/rs/981-LQX-968/images/AR-SA-2022-16th-Annual-State-Of-Agile-Report.pdf
4. Scrum Alliance. (2018). State of Scrum 2017-2018. scrumalliance.org. Retrieved April 17, 2023, from https://www.scrumalliance.org/ScrumRedesignDEVSite/media/

ScrumAllianceMedia/Files%20and%20PDFs/State%20of%20S
crum/2017-SoSR-Final-Version-(Pages).pdf

5. Sutherland, J., & Schwaber, K. (2020). Scrum Guides, 5. Scrum Guide. Retrieved April 7, 2023, from https://scrumguides.org/scrum-guide.html

6. Sutherland, J., & Schwaber, K. (2020). Scrum Guides, 7-8. Scrum Guide. Retrieved April 7, 2023, from https://scrumguides.org/scrum-guide.html

7. Sutherland, J., & Schwaber, K. (2020). Scrum Guides, 3-4. Scrum Guide. Retrieved April 7, 2023, from https://scrumguides.org/scrum-guide.html

8. Sutherland, J., & Schwaber, K. (2020). Scrum Guides, 5-7. Scrum Guide. Retrieved April 7, 2023, from https://scrumguides.org/scrum-guide.html

9. Horowitz, B. (2012, June 15). Good Product Manager/Bad Product Manager. Andreessen Horowitz. Retrieved April 17, 2023, from https://a16z.com/2012/06/15/good-product-managerbad-product-manager/

10. Sutherland, J., & Schwaber, K. (2020). Scrum Guides, 7-8. Scrum Guide. Retrieved April 7, 2023, from https://scrumguides.org/scrum-guide.html

11. Sutherland, J., & Schwaber, K. (2020). Scrum Guides, 12. Scrum Guide. Retrieved April 7, 2023, from https://scrumguides.org/scrum-guide.html

12. Sutherland, J., & Schwaber, K. (2020). Scrum Guides, 10-11. Scrum Guide. Retrieved April 7, 2023, from https://scrumguides.org/scrum-guide.html

13. Sutherland, J., & Schwaber, K. (2020). Scrum Guides, 11. Scrum Guide. Retrieved April 7, 2023, from https://scrumguides.org/scrum-guide.html

14. Sutherland, J., & Schwaber, K. (2020). Scrum Guides, 9. Scrum Guide. Retrieved April 7, 2023, from https://scrumguides.org/scrum-guide.html

15. Sheridan, R. (2015). Joy, Inc.: How We Built a Workplace People Love, 65-67. Penguin Publishing Group.

16. Sutherland, J., & Schwaber, K. (2020). Scrum Guides, 8-9. Scrum Guide. Retrieved April 7, 2023, from https://scrumguides.org/scrum-guide.html

17. Agile Alliance. (2017, June 13). Backlog Refinement. agilealliance.org. Retrieved April 17, 2023, from https://www.agilealliance.org/glossary/backlog-refinement/

18. Sutherland, J., & Schwaber, K. (2020). Scrum Guides, 9. Scrum Guide. Retrieved April 7, 2023, from https://scrumguides.org/scrum-guide.html
19. Sutherland, J., & Schwaber, K. (2020). Scrum Guides, 10. Scrum Guide. Retrieved April 7, 2023, from https://scrumguides.org/scrum-guide.html
20. Hohmann, L. (2007). Innovation Games: Creating Breakthrough Products Through Collaborative Play. Addison-Wesley.
21. Leal, C. J. (2022, December 9). Little's Law. Kanban Zone. Retrieved April 18, 2023, from https://kanbanzone.com/resources/lean/littles-law/
22. Agile Alliance. (2016, April 5). What is Velocity in Agile? Agile Alliance. Retrieved April 18, 2023, from https://www.agilealliance.org/glossary/velocity/
23. Cohn, M. (2004). User Stories Applied, 117-120. Addison-Wesley.
24. Hoory, L., & Bottorff, C. (2022, March 27). What Is A Burndown Chart? – Forbes Advisor. Forbes. https://www.forbes.com/advisor/business/what-is-a-burndown-chart/
25. Cohn, M. (2004). User Stories Applied, 121-123. Addison-Wesley.
26. Cohn, M. (2004). User Stories Applied, 88. Addison-Wesley.
27. Lithespeed. (n.d.). The Throw Down: Agile Estimation vs. #NoEstimates. Lithespeed. Retrieved April 19, 2023, from https://lithespeed.com/throw-agile-estimation-vs-noestimates/
28. Cohn, M. (2005). Agile Estimating and Planning. Prentice Hall Professional Technical Reference.
29. Cohn, M. (n.d.). Planning Poker: An Agile Estimating and Planning Technique. Mountain Goat Software. Retrieved May 16, 2023, from https://www.mountaingoatsoftware.com/agile/planning-poker
30. Cagan, M. (2017). INSPIRED: How to Create Tech Products Customers Love. Wiley.
31. Cagan, M., & Jones, C. (2020). EMPOWERED: Ordinary People, Extraordinary Products (pp 288-291). Wiley.

Chapter 6:
1. Rasure, E. (2022, September 28). What Is the Kanban System? Investopedia. Retrieved April 19, 2023, from https://www.investopedia.com/terms/k/kanban.asp#
2. Anderson, D. J. (2010). Kanban: Successful Evolutionary Change for Your Technology Business. Blue Hole Press.

3. Leal, C. J. (2022, December 9). Little's Law. Kanban Zone. Retrieved April 18, 2023, from https://kanbanzone.com/resources/lean/littles-law/
4. Kanbanize. (n.d.). Introduction to Kanban Metrics and Reporting. Kanbanize. Retrieved April 19, 2023, from https://kanbanize.com/kanban-resources/kanban-analytics
5. Ladas, C. (2021, July 30). What is Scrumban? Agile Alliance. Retrieved April 19, 2023, from https://www.agilealliance.org/scrumban/

Chapter 7:
1. Agile Alliance. (2016, July 13). Pair Programming: Does It Really Work? Agile Alliance. Retrieved April 19, 2023, from https://www.agilealliance.org/glossary/pairing/
2. Sheridan, R. (2015). Joy, Inc.: How We Built a Workplace People Love, 50-54. Penguin Publishing Group.
3. Agile Alliance. (n.d.). What is Mob Programming? Agile Alliance. Retrieved April 19, 2023, from https://www.agilealliance.org/glossary/mob-programming/
4. Alistair, C., & Laurie, W. (2000, February). The Costs and Benefits of Pair Programming. Laurie Williams. Retrieved April 19, 2023, from https://collaboration.csc.ncsu.edu/laurie/Papers/XPSardinia.PDF
5. Jeffries, R., Williams, L., Robert, K., & Ward, C. (2000). Strengthening the Case for Pair Programming. Laurie Williams. Retrieved April 19, 2023, from https://collaboration.csc.ncsu.edu/laurie/Papers/ieeeSoftware.PDF
6. Thanks to Singlewire Software Senior Developer & Engineering Manager, Sarah Murphy, for her thoughts on trying Pair Programming, April 2023.
7. Thanks to Singlewire Software Senior Developer & Author, James Elliott for his thoughts on concentration during Pair Programming, April 2023.
8. Zettler, K. (n.d.). Trunk-based Development. Atlassian. Retrieved April 25, 2023, from https://www.atlassian.com/continuous-delivery/continuous-integration/trunk-based-development
9. Schiestl, B. (2020, February 28). What Is a Branch? | Perforce. Perforce Software. Retrieved April 25, 2023, from https://www.perforce.com/blog/vcs/branching-definition-what-branch#

10. GitHub. (n.d.). Resolving a merge conflict using the command line. GitHub Docs. Retrieved April 25, 2023, from https://docs.github.com/en/pull-requests/collaborating-with-pull-requests/addressing-merge-conflicts/resolving-a-merge-conflict-using-the-command-line

11. Fowler, M. (2020, May 7). FeatureBranch. Martin Fowler. Retrieved April 25, 2023, from https://martinfowler.com/bliki/FeatureBranch.html

12. Atlassian. (n.d.). Gitflow Workflow. Atlassian. Retrieved April 25, 2023, from https://www.atlassian.com/git/tutorials/comparing-workflows/gitflow-workflow

13. Zettler, K. (n.d.). Trunk-based Development. Atlassian. Retrieved April 25, 2023, from https://www.atlassian.com/continuous-delivery/continuous-integration/trunk-based-development

14. Thanks to Singlewire Software Director of Software Development, Charlie Ciccia, for his great summary of why pushing a lot of code at once is a very bad idea, April 2023.

15. Thanks to Singlewire Software Senior Developer & Author, James Elliott for pointing out that engineers can't always dictate how long their releases are, April 2023.

16. Thanks to Singlewire Software Senior Developer & Engineering Manager, Sarah Murphy, all the great survival tips for the Branching & Merging section, April 2023.

17. Rinaldi, B. (n.d.). Feature Flags—What Are Those? Uses, Benefits & Best Practices. LaunchDarkly. Retrieved April 25, 2023, from https://launchdarkly.com/blog/what-are-feature-flags/

18. Sato, D. (2014, June 25). CanaryRelease. Martin Fowler. Retrieved April 25, 2023, from https://martinfowler.com/bliki/CanaryRelease.html

19. Aston, B. (2023, March 10). 10 Best Feature Flag Software For Managing Feature Flags [2023]. The Product Manager. Retrieved April 25, 2023, from https://theproductmanager.com/tools/best-feature-flag-software/

20. Ambler, S. (n.d.). Introduction to Test Driven Development (TDD). Agile Data. Retrieved April 20, 2023, from http://agiledata.org/essays/tdd.html

21. Beck, K. (2003). Test Driven Development. Addison-Wesley.

22. Debois, P., Kim, G., Willis, J., Forsgren, N., & Humble, J. (2021). The DevOps Handbook, Second Edition: How to Create World-

Class Speed, Reliability, and Security in Technology Organizations. IT Revolution Press.
23. Kniberg, H., & Ivarsson, A. (2012, October). Scaling Agile @ Spotify. Crisp's Blog. Retrieved April 20, 2023, from https://blog.crisp.se/wp-content/uploads/2012/11/SpotifyScaling.pdf
24. Knaster, R. (n.d.). SAFe 6.0. Retrieved April 20, 2023, from https://scaledagileframework.com/
25. The LeSS Company B.V. (n.d.). LeSS. Overview - Large Scale Scrum (LeSS). Retrieved April 20, 2023, from https://less.works/
26. Ambler, S. W., & Lines, M. (2012). Disciplined Agile Delivery: A Practitioner's Guide to Agile Software Delivery in the Enterprise. IBM Press.
27. Kniberg, H., & Ivarsson, A. (2012, October). Scaling Agile @ Spotify. Crisp's Blog. Retrieved April 20, 2023, from https://blog.crisp.se/wp-content/uploads/2012/11/SpotifyScaling.pdf
28. Kniberg, H. (2014, March 27). Spotify engineering culture (part 1) - Spotify Engineering. Spotify Engineering. Retrieved April 20, 2023, from https://engineering.atspotify.com/2014/03/spotify-engineering-culture-part-1/
29. Kniberg, H. (2014, September 20). Spotify engineering culture (part 2) - Spotify Engineering. Spotify Engineering. Retrieved April 20, 2023, from https://engineering.atspotify.com/2014/09/spotify-engineering-culture-part-2/
30. Kniberg, H. (2014, September 16). Squad Health Check model - visualizing what to improve - Spotify Engineering. Spotify Engineering. Retrieved April 20, 2023, from https://engineering.atspotify.com/2014/09/squad-health-check-model/
31. Business Agility Institute. (n.d.). Business Agility Institute. Business Agility Institute. Retrieved April 20, 2023, from https://businessagility.institute/

Other Recommended Sites:

Lithespeed (https://lithespeed.com/) is the wonderful agile consulting partner and training organization that worked with me to introduce agile software development to National Geographic. I've worked with Sanjiv Augustine, Bob Payne, and a number of their coaches many times. If you're looking for training or an amazing agile partner, I highly recommend them.

I am currently the Vice President of Software Development at Singlewire Software (https://www.singlewire.com/). This is a shameless plug for our emergency management, mass notification, and visitor management platforms. We keep kids safe at school, doctors and patients safe in hospitals, and do the same for many, many other types of organizations.

UW-Madison (https://www.cs.wisc.edu/) is my alma mater and where I teach the Computer Science Capstone course. If you're a student there, I'd love to see you in class! If you're a company, here comes another shameless plug for you to partner with the course and meet some of the great students at UW. You can reach out to me here: https://www.cs.wisc.edu/staff/field-amber/.

Image Credits

1. Introduction: Photo likely taken by Julius Laiser, August 2008. IBM Corporate Service Corps Tanzania Team 1. Enduimet Wildlife Area, Tanzania.
2. Chapter 1: The Waterfall Process image was created by Graphic Designer, Emily Varone, May 2023. https://emilyvarone.com/about.
3. Chapter 1 & Part II: The Agile Umbrella image was created by Amber Field, 2017.
4. Chapter 2: The Self-Selection Process was created by Nomad8 & Sandy Mamoli, 2016. Permission granted for use by David Mole. https://nomad8.com/articles/self-selection-on-a-page.
5. Chapter 2: Self-Selection Survey data developed by Amber Field & Jesse Huth, 2016.
6. Chapter 3: The Scientific Process image was created by Graphic Designer, Emily Varone, May 2023. https://emilyvarone.com/about.
7. Chapter 3: The Build, Measure, Learn image was created by Graphic Designer, Emily Varone, May 2023. https://emilyvarone.com/about.
8. Chapter 3: Ways to Test slide courtesy of Charles Moore, Capital One Labs, 2017. Charles is now an Executive Coach and Strategy Consultant at Thrive Street Advisors (https://www.thrivestreetadvisors.com/).
9. Chapter 3: Capstone Class Wallet Project photo taken by Amber Field, October 2021.
10. Chapter 3: Black & White InformaCast wireframe created by Singlewire Software UX Lead, Ben Koca, 2021. https://www.benkoca.com/.
11. Chapter 3: Stanford d.school streamlined design process (Legacy, circa 2012) used courtesy of Stanford d.school's Shree Vijayaraj.
12. Chapter 3: Google Design Sprint image used courtesy of Jake Knapp, Google Ventures. Permission granted for use by Jake Knapp & Sylvie Carr, May 2023.

13. Chapter 4: Screen shot of Amazon's Home Page taken by Amber Field on May 17, 2022.
14. Chapter 4 & 5: The Product Backlog image was created by Graphic Designer, Emily Varone, May 2023. https://emilyvarone.com/about.
15. Chapter 5: The Scrum Process image was created by Graphic Designer, Emily Varone, May 2023. https://emilyvarone.com/about.
16. Chapter 5: The Scrum Team image was created by Graphic Designer, Emily Varone, May 2023. https://emilyvarone.com/about.
17. Chapter 5: The Sprint Backlog image was created by Amber Field, May 2023.
18. Chapter 5: The Product Backlog mock-up in JIRA was created by Amber Field, Fall 2020.
19. Chapter 5: The Sprint Backlog in JIRA is a screenshot of a Capital One Labs team backlog created by Amber Field, January 2017.
20. Chapter 5 & 6: A Kanban-Style Team Board in JIRA is a screenshot of a Singlewire team board created by Amber Field, 2021.
21. Chapter 5: A Google Document Retrospective is a screenshot of a Singlewire team retrospective created by Amber Field, 2019.
22. Chapter 5: Sailboat Retrospective Example - Contributed to https://openpracticelibrary.com/practice/retrospectives/ by Tim Beattie and Matt Takane and originally based on Luke Hohmann's "Speed Boat", November 2017. Used under Creative Commons Attribution 4.0.
23. Chapter 5: Sample Sprint Calendar was created by Amber Field, Fall 2020.
24. Chapter 5: Example Velocity Chart was created by a Capital One Labs team in JIRA. Screenshot taken by Amber Field, 2017.
25. Chapter 5: Sample Burndown Chart courtesy of Pablo Straub via Wikimedia Commons (https://commons.wikimedia.org/wiki/File:SampleBurndownChart.png#filelinks), published June 26, 2009.
26. Chapter 5: Bad Burndown Chart courtesy of Jacque Harper (https://agilecoachjacque.wordpress.com/2014/11/18/over-the-past-year-or-so-ive-collected-some-burndown-charts-that-illustrate-some-team-problems/), November 2014.
27. Chapter 7: Chapter 5: Gitflow image was created by Graphic Designer, Emily Varone, May 2023. https://emilyvarone.com/about.

28. Chapter 7: Chapter 5: Trunk-based Deployments image was created by Graphic Designer, Emily Varone, May 2023. https://emilyvarone.com/about.
29. Chapter 7: DevOps image used courtesy of Muhammad Ribkhan on vecteezy.com, No Date.
30. Chapter 7: Spotify Team Model image courtesy of Henrik Kniberg (https://blog.crisp.se/wp-content/uploads/2012/11/SpotifyScaling.pdf), October 2012.
31. About the Author: Photo taken by Nathan King, October 2015.

Index

About The Author

Amber Field is the Vice President of Software Development at Singlewire Software and teaches the Computer Science Capstone course at the University of Wisconsin-Madison. She has been an engineer, agile coach, and leader at a number of organizations including IBM, National Geographic, Oracle Utilities (formerly Opower), and Capital One Labs. Amber frequently speaks at conferences and blogs at amberrfield.com. She lives in Madison, WI.

www.ingramcontent.com/pod-product-compliance
Lightning Source LLC
Chambersburg PA
CBHW071417210326
41597CB00020B/3552